George Murray Humphry

The Human Foot and the Human Hand

George Murray Humphry

The Human Foot and the Human Hand

ISBN/EAN: 9783337371043

Printed in Europe, USA, Canada, Australia, Japan

Cover: Foto ©berggeist007 / pixelio.de

More available books at **www.hansebooks.com**

AND THE

HUMAN HAND.

BY

G. M. HUMPHRY, M.D. F.R.S.

LECTURER ON ANATOMY AND PHYSIOLOGY IN THE
UNIVERSITY OF CAMBRIDGE.

MACMILLAN AND CO.

Cambridge:

AND 23, HENRIETTA STREET, COVENT GARDEN,

London.

1861.

THE following pages originated in two popular Lectures which were delivered in Cambridge. In the preparation for publication many additions have been made; but I have thought it best to retain the original form.

CONTENTS.

THE HUMAN FOOT.

THE HUMAN HAND.

THE HUMAN FOOT.

THE Human Body is one of the most worthy objects of man's study. It is the noblest as well as the crowning work of creation. In it material organization is carried to the greatest perfection. It surpasses, therefore, all other physical objects in exquisiteness of construction and in interest. How comes it, then, that most persons are so ignorant respecting it? Men, well informed in other matters, are usually altogether uninformed with regard to this. In every other branch of science we find amateur students pursuing the subject with zeal and success. Geology, Chemistry, Botany, Zoology, and even Comparative Anatomy have each their votaries; but Human Anatomy attracts no one. Why is this? Partly, I think, because opportunities for acquir-

H.   1

ing such information as is suitable and interesting
are not so many as they ought to be.

It must be confessed, also, that we teachers of
Anatomy are somewhat to blame. We are too
prone, in our Lectures and Examinations, to dwell
upon bare details, without enlivening those details with the many bright features of interest with
which they are naturally invested; and we fail,
therefore, to render it so attractive a science as it
might be. The example of those able and animated
teachers, John and Charles Bell, who laboured
with some success to disperse the clouds that
have ever overhung the horizon of anatomy, has
been too much forgotten; and the flame which
they kindled has almost died out under the chilling apathy of their successors. Truly glad should
I be to see a change in this. I cannot but think
that if the teachers of Anatomy took higher
and more philosophical views of their science
there would be no lack of interest on the part
of the students. The interest so excited would
soon spread beyond the limits of the profession;
and there would thus be opened up to the public
some of the products of that rich vein of knowledge and of that abundant material for thought
which lie buried in the human frame.

I therefore willingly accede to your request for
a Lecture upon some part of the anatomy of the

human body, relying upon the intrinsic interest of
the subject to make amends for my own deficiencies
in expounding it; and I select the HUMAN FOOT,
because a few of the more important points of its
construction can be explained without much difficulty, because it affords a good illustration of
some of the principles of animal mechanism, and
because its form constitutes one of the great
characteristics whereby man is distinguished from
the lower animals. As an instrument of support
and of locomotion it excels the foot of any other
animal. It evinces its excellence by enabling
man to stand upright in a way that no other
animal can do; and so efficiently does the foot
accomplish this and perform the task of carrying
the body, that the hand is set at liberty to
minister to the will. Thus is the foot instrument-
al in giving us an advantage over other animals,
and in enabling us to provide the means of de-
fence; and, thus, it aids us to carry out those
wondrous works which are second only to the
marvellous results of creative power.

We are accustomed to regard the hand as the
great agent by which all this is attained, and
we are apt to forget how much it is indebted
to the foot. We do not reflect that, if the foot
of man presented no distinguishing peculiarity,
the hand, like the corresponding part in other

animals, would be compelled to share with it the task of carrying the body, and could, therefore, not be devoted to the various offices which it is now free to perform. Little right has the hand to say to the foot, "I have no need of thee."

The principle of "division of labour."

In this concentration of locomotive power in the foot we have an illustration of what is called the "principle of division of labour," a principle with which all civilized communities are familiar, and to which we are much indebted for the present advanced state of the arts and sciences; but which we may be said to have borrowed from the economy of nature. We find ever-increasing manifestations of it as we ascend in the animal series, from the lower and more simple to the higher and more complicated forms. Indeed, just as each step in civilization is attended with a further development of this principle, so is each division of the animal kingdom distinguished from those below it by the more distinct assignation of particular functions to particular organs, and by the consequent improvement of the mode in which the functions are performed. While, in proportion as the several organs acquire more distinct speciality in their work, so do they become, more and more, dependent upon one another, and, more and more, subjected to the control of central government, which is represented by the brain.

For instance, some of the lower animals, as the fresh-water POLYP, present nearly a uniform structure throughout their whole substance; and every part of them consequently performs the same function. There is not one organ for digestion, another for circulation, a third for respiration, and so on; but all these functions are performed by the same structure, and are performed, therefore, in a rude and imperfect manner. Any portion of the creature possesses all the requisites for its own nutrition, and is, so, independent of the remainder, and can live alone. Hence, the polyp may be divided into a number of pieces, each of which goes on living. Gradually, as we ascend from these lowly beings to the higher classes of animals, we find organs and functions more and more distinct from another; a Stomach is provided for the work of digestion, a Heart for circulation, Lungs for respiration. Each of these organs is essential to the existence of the others and of every part of the body; and they are all maintained in harmonious co-operation by the presiding influence of the nervous system.

Or, trace one of the *functions* in illustration of the same principle. Take the function of Locomotion, which has an especial relation to our present subject. In the LEECH and the WORM the whole length of the body is occupied in the work, one part as much as another; and still, it is but a crawl. In the FISH the whole body is buoyed up by the water; it is flattened from side to side, and is all, from the head to the tail, concerned in the lateral stroke by which the animal is driven along; the side fins, which are the representatives of limbs, doing little beyond serving to guide and balance. In the other VERTEBRATES the work of locomotion is so far concentrated as to be assigned, almost entirely, to the limbs. All *four* limbs are in most of them devoted to it; while the bones and muscles of the trunk are only indirectly concerned in it. In MAN, however, *two* limbs only are assigned to this important office. In him, therefore, the concentration of locomotive power, in other words the principle of division of labour, is carried out to the greatest extent—a disposition which affords one of the many proofs that the construction of his body combines with the faculties of his mind to place him at the head of the animal kingdom.

In making comparisons of different animals

with one another, and in speaking of the relative perfection of their several organs, we must not forget that *every* organ of every animal is perfect as regards the purpose for which it was made. But some animals are said to occupy a higher position than others, or to be superior to others, because their mechanism is more complex, and they are, thereby, enabled to perform a greater variety of functions. And, in the animal kingdom, in proportion as each function rises into prominence, and becomes well and distinctly performed, so is a special organ assigned to it, and that organ becomes more and more highly elaborated.

You will not misunderstand me, then, when I say that concentration of function and perfection of structure usually go together. And, forasmuch as in the lower limbs of man there is a greater concentration of locomotive function than in any other part of any other animal, you will expect to find, in them, a greater perfection of locomotive mechanism—that is to say, a more complete combination of strength with variety, rapidity, and extent of movement—than is elsewhere to be met with.

This consideration will ensure attention while I give a brief account of the anatomy of man's lower limb, more particularly of the foot.

Structure of the Lower Limbs.

The weight of the trunk is transmitted to the knee (see fig. 4, p. 15) by a single bone—the thigh-bone. This is the longest bone in the body, measuring, on the average, nearly eighteen inches. Above, it is jointed with the haunch-bone of the pelvis at the hip-joint. From the knee two bones descend to the ankle. Of these one is much the larger, and bears the chief of the weight. The other serves to give attachment to muscles, and to strengthen the ankle-joint. It runs down on the outer side of the ankle, forming there what is called the "outer ankle;" and a process of the larger bone runs down, in like manner, on the inner side, and forms the "inner ankle." The front and inner side of the larger bone are close under the skin. This part is called the "*shin*," being so named perhaps from the word "chine" or edge, because the leg presents an edge along the front, to facilitate its cleaving a way through the air, water, grass, or underwood. The shin itself is not particularly tender; but the skin is a good deal exposed here, and, as it lies so near the hard bone, it is easily injured; and, when "broken," it is often difficult to heal.

In some very tall persons, and particularly in those who are so tall as to be called GIANTS, I have found the leg or shank bones, that is, the bones between the knee and the ankle, very long, disproportionately long to the rest of the skeleton. They are so in the skeleton of the Irish Giant, O'Byrne, which is preserved in the Museum of the College of Surgeons, in another Irish Giant in the Museum of Trinity College, Dublin, and in some other specimens which I have had an opportunity of measuring. In the name "Long Shanks" given to Edward I., the word "shanks" probably included the thigh as well as the leg, just as we are in the habit of applying the word "leg" to the whole of the lower limb.

Bones of the Foot.

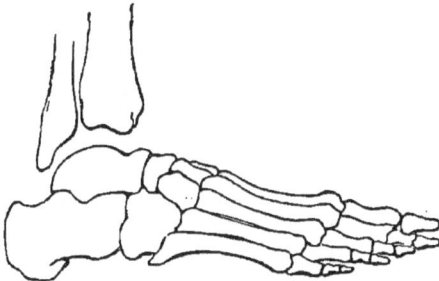

Fig. 1.

There are 26 bones in the FOOT. The hinder 7—called *tarsal* bones—are short and thick; they form the hinder part of the instep. In front of them lie 5 *metatarsal* bones, one passing, forwards, from the fore part of the tarsus to each toe. Behind, these are close together, and are connected with the tarsus. As they run forwards they diverge a little from one another; and their anterior ends rest upon the ground, and form the "balls" of the toes. They constitute the fore part of the instep. The remaining 14 bones are the toes. They are arranged in rows, like soldiers in a phalanx, three deep, and are hence called *phalanges*.

You observe that, although each of the other toes has 3 bones, the great toe has only 2. In this respect, therefore, it is an imperfect, or, rather, an incomplete member. The deficiency does not depend upon a want of length in the great toe; for this is usually as long as the second toe; in some persons it is a good deal longer; and it is always distinctly longer than the outer two toes. The reason for there being only two phalanges instead of three probably is because the great toe is required to be stronger than any of the others; and an additional bone would have tended to weaken it. I have, elsewhere[1], given reasons for thinking that it is the middle phalanx which is absent in the great toe.

---

[1] *Treatise on the Human Skeleton*, p. 395.

It is a curious and interesting fact, affording a remarkable illustration of the close adherence to a uniform plan which has been observed in the construction of the various animals, that, in no instance, does this toe contain more than two bones. Even in those creatures, as the SEAL (fig. 2), in which it attains to greater length than any of the other

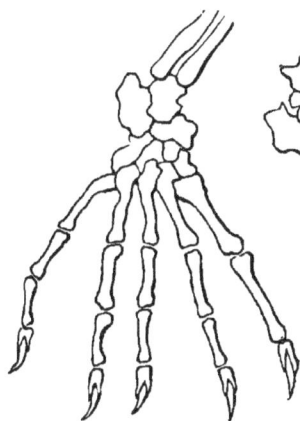

Fig. 2.
Seal's Foot.

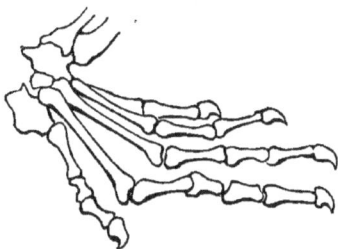

Fig. 3.
Lizard's Foot.

sprawling digits, it contains the same number of bones as in man, its extraordinary length being attained by an elongation of the two bones, not by the addition of a third. And in those animals, as certain Lizards (fig. 3), where the number of bones in the other toes is increased to 4 or even 5, the number in the first, or inner, toe is still no more than two. The same rule applies to the fore limb; the number of bones in the inner digit, which, in

man and monkeys, is called the "thumb," is in no
case more than two. In some animals, as will be
mentioned again, there is only one bone in this
digit, and in some the digit is wanting altogether;
but in none does it contain *more* than *two* bones.

This reminds me of a still more remarkable instance of adherence to a particular number of bones.
In the mammalian group of animals the *neck*, with
only one or two exceptions, contains *seven* bones,
neither more nor less. Whether it be the long
neck of the GIRAFFE, or the short neck of the
MOUSE, the BAT, or the PORPOISE, each consists,
like the neck in MAN, of seven bones. For what
reason a particular number should be thus rigidly
observed, it is not easy to say.

Of the seven tarsal bones the uppermost (fig. 1)
is called the *astragalus*, from a supposed resemblance to a die. It is the middle bone of the
instep. Above, it is jointed with the leg-bones;
behind, it is connected with, and rests upon, the
heel-bone, which is the largest bone in the foot.
The bone which lies immediately in front of the
astragalus, and supports it in this direction, is
called the *scaphoid*, or boat-like, bone. In front of
it are three *wedge-bones*, each of which is connected
with one of the metatarsal bones of the inner
three toes. On the outer side of the wedge-bones,
connected with the metatarsals of the two small

toes, and locked in between them and the heel-bone, is the *cuboid* bone.

I must confine my remarks chiefly to the *human* foot. Still the anatomy of man derives so much interest from being studied in connexion with that of the lower animals, and is so much more instructive when this is done, that I cannot forbear diverging, here and there, to make a few comparisons. Let me, for a moment, draw your attention to a similarity, in general construction, which exists between the lower limbs of man, and the hinder limbs of other animals. And the comparison may be extended to the fore limbs; for however diverse may be the appearance and the mode of action of the limbs in different animals, whether they be terminated by hands or by feet, whether they move upon the ground or ply in air or water, whether they be attached to the head, as are the front fins in many fishes, or, as is more common, be situated at the fore and hinder parts of the trunk, the same plan is traceable in all.

Great, indeed, is the variety of detail in nature. It is everywhere observable. No two things, however near their resemblance, are precisely alike. Yet, as I have before said, there is a remarkable adherence to unity of plan. One star differs from another star in glory, yet all appear fashioned in the same manner, and subject to the same laws. There are almost infinite varieties in the vertebrate kingdom. Each animal exhibits its own peculiarities; yet they are all formed in the same manner, and are developed upon one fundamental pattern, diverging from it in different ways according to the requirements of each. Again, though the several parts of the same animal differ from one another; yet in the skeleton the same bones which exist in one part may, as a general rule, be traced in other parts and in other animals. The bones which make up the pelvis in man are repeated in his shoulder, and, even, in his skull; and they may be recognised in the pelvis, in the shoulder, and in the skull, of all other vertebrate animals, with few exceptions. They undergo, it is true, great varieties in shape and size; but they can be shown to be the same, or, in the language of anatomists, to be "homologous." It is highly interesting to the anatomist to trace the same bone through the different parts of the same animal, and through the various animals of the vertebrate series, and to observe the modifications which it undergoes in order to adapt it to the multiform mechanism of the several classes, to observe it sometimes dwindling, or even vanishing, and then, it may be, reappearing under some new conditions.

I must, however, resist the temptation to wander into this attractive field. It will suffice to

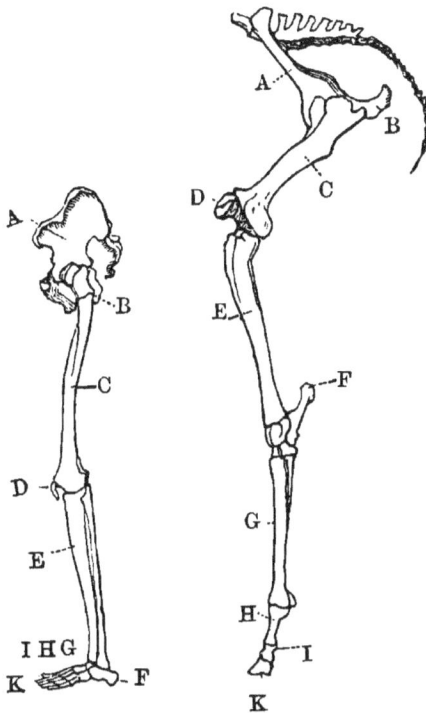

Fig. 4.   Fig. 5.
Human Leg.   Horse's Leg.

take an illustration by a comparison of the bones
of the human lower limb with those of the hind
limb of the Horse. This may be easily done by
the aid of these drawings (figs. 4 and 5) in which
the two limbs are placed side by side, and the corresponding bones are marked with the same letters. Notwithstanding the many points of difference the same plan will be recognised in each.
There is in each the thigh (c), the leg (E), and the
foot, with the tarsal and metatarsal (G) bones, and

the phalanges (H, I, K). But in the HORSE two of the digits (the marginal ones, that is, the great toe and the little toe) are wanting, two are rudimentary, and the remaining one, which corresponds with the middle toe of man, in length, size, and strength, more than makes amends for the deficiency of the others. The lowermost bone, or terminal phalanx, of this huge toe, called the *coffin-bone* (K), is encased in the hoof, which corresponds with the human nail, and is the only part of the foot that rests upon the ground.

In MAN the whole weight of the body has to be borne upon *two* feet; often it is balanced upon *one*. The foot is, consequently, spread out; and all the bones, from the heel to the tips of the toes, are made to form the basis of support upon the ground. The HORSE, on the contrary, having no hands, but *four* feet, does not require so great breadth in each foot; and the opportunity is taken to narrow the foot, and to lengthen it so as to give fleetness. The end is attained by suppressing some of the toes, by elongating one far beyond the others, and enduing it with such strength as to enable it to carry the requisite weight upon the tip of the last phalanx. The heel (F) is raised high above the ground and becomes the "hock." To speak of a horse *kicking with his heels* is, therefore, about as correct as to say, that he *breaks his knees.* His knee, as you perceive by the position of the "knee-cap" (D), is high up in the hind limb, near his body, quite out of harm's way in a fall. The fact is, that he kicks with his *toes;* and, when he falls, he cuts the skin over the part in his *fore* limbs, which corresponds with the back of our *wrists.*

In the upper segment, or thigh, the difference between the two limbs is seen to be, to a certain extent, the reverse of what it is below. That is to say, whereas, in the HORSE, the *toe* is elongated and thickened, so as greatly to exceed the corresponding part of the human limb; in MAN the *thigh-bone* is elongated, so as to be double the length of that of the horse; the thigh-bone in man is also placed more vertically, nearly in the plane of gravity of the trunk. The horse's thigh-bone slants forwards and outwards, which gives the muscles great power by causing them to run more at right angles between their points of attachment; and this arrangement increases the strength of the animal in drawing weights, and facilitates springing. A man cannot spring without first bending the limbs a little; whereas a horse, or a goat, can spring, at once, from the position in which it is standing.

To revert to the anatomy of the Human Foot.

H.   2

The Arch of the Foot.

The seven tarsal and the five metatarsal bones
—that is, the twelve bones of the instep—are arranged and jointed together so as to form an arch
from the point of the heel to the balls of the toes.
This is called the "plantar arch," from the Latin
word *planta*, the sole of the foot. The *astragalus*
forms the summit, or key-bone, of the arch. It
receives the weight from the leg, and transmits it,
through the hinder pillar of the arch, to the heel,
and, through the front pillar of the arch, to the
balls of the toes.

The drawing represents a section, from behind
forwards, of the lower end of the leg-bone, and of

Fig. 6.

the bones lying along the inner side of the plantar

arch. Behind it extends through the heel-bone, and in front through the great toe. It exhibits the arrangement of the fibres and plates in the interior of the bones, and shows that the greater number of them, in each bone, follow the direction of the two pillars of the arch; that is to say, they descend from the summit of the arch where it supports the leg-bone, backwards and downwards, to the heel, and, forwards and downwards, to the balls of the toes. Their arrangement is, therefore, such as to give resisting strength to the bones in the directions in which it is most required.

You may think that the arch of the foot would have been a much simpler structure, as well as stronger, if it had been composed of one bone instead of several. But it must be remembered that it would, then, have been liable to be cracked and broken by the sudden and violent manner in which, during running and jumping, the weight of the body is thrown upon it. Moreover, the several bones, where they touch one another, are covered with a tolerably thick layer of highly elastic gristle or cartilage (represented by the clear line left in the drawing along the contiguous edges of the bones); and this provision, together with the slight movements which take place between these bones, gives an elasticity to the foot and to the step, and serves to break the jars and shocks which are caused by the sudden contact of the foot with the ground.

This last is a very important point; and we find numerous contrivances in different parts of the body to protect the brain and other delicate organs from jars. So efficient are these contrivances, and so exact is the adaptation of the mechanism of the limbs and the trunk to the texture of the internal organs, that, while these are in a healthy state, we are able to run, to jump, and to leap from a considerable height, without inconvenience. But, if the organs be inflamed, or if the nervous system be over sensitive, as in common headache, the provisions, which are calculated for the normal state, are insufficient; ordinary movements are then painful, and to jump is intolerable.

The muscles play a very essential part in this work. *First,* they place the limbs in the most favourable position. Thus, when we alight upon the ground, from a height, we always contrive to do so with the knees and hips a little bent, so that the limbs readily yield at the joints, and act as springs to break the jar. Elderly persons commonly keep the limbs bent, even when walking quietly along. They do this because they need all the benefit which position will afford to make amends for the loss of elasticity consequent on the thinning and drying of the cartilages, and other changes that take place in the body with advancing years. *Secondly*, the muscles brace the limbs and joints in the position in which they have placed them. We experience the effect of the want of this salutary influence when we kick against an unseen object, or fall suddenly, or receive any blow or shock for which we are unprepared. How disagreeable, to say the least, it is to make the step for an additional stair when we have arrived at the top of a staircase, or, still worse, to meet with an unseen stair when we think that we have got to the bottom.

You perceive from the drawing (fig. 6) that there is a great difference between the two pillars of the plantar arch. The hinder pillar is comparatively short, and narrow, and descends suddenly, almost in a vertical direction, from the ankle, to the ground; and it is composed of only one bone— the heel-bone—which is jointed directly with the astragalus: whereas the fore pillar is longer and broader, is composed of several bones jointed together, and slopes much more gradually to the ground. There is, therefore, far less elasticity in the hinder part of the foot than in the fore part. Hence, when we descend from a height upon the ground, we always alight upon the balls of the toes, and thus gain the advantage which the several bones and joints afford in breaking the shock. If, after going up stairs this evening, you take the

trouble to come down again, you will find that
you alight upon each stair on the balls of the toes
and experience no inconvenience, however quickly
the descent is made. But, if you change the mode
of proceeding, and descend upon the heels, the
feeling will be by no means agreeable; and the
various organs of the body, being disturbed from
their accustomed repose, will raise such remon-
strances against your infringement upon nature's
ways, that you will scarcely be able to continue
the experiment. Proportionately more distressing
is the sensation caused by jumping from a chair
upon the heels. Indeed, this is not done altogether
without risk; and the trial of it is scarcely to be
recommended to persons who have attained to that
sober period of life at which we are willing to con-
cede that, in some things, nature is wiser than
ourselves. Only a short time since I saw a gentleman, who, in jumping down some steps into a
back yard, accidentally came upon his heels, and
jarred one hip so severely that he was confined to
his sofa for several days in consequence.

But, you may say, " in walking we do place the
heel upon the ground first and experience no inconvenience." True, 'because the force with which
the foot descends in walking is very slight; and
the weight is directed upon the heel, obliquely, in
such a manner as to bring the toes very quickly to

the ground, and really to throw nearly the whole force in that direction. Moreover, you may observe that when we walk, the weight of the body is partly sustained by the fore part of the one foot till the whole of the other foot is on the ground. I will, however, revert to the disposition of the feet in walking and running presently.

The arch of the foot has to bear great weight and at great disadvantage; and there is very little in the *shape* of the bones to maintain its integrity. Indeed, they all fall asunder when the other structures are removed, the key-bone dropping through of its own weight. And the same thing may be remarked throughout the skeleton. Wherever two or more bones move upon one another, their surfaces are so constructed that they do not hold together without some assistance from the soft parts. There are joints in the body which we call "hinge-joints," and others which we call "ball-and-socket joints;" but in none of them is there such a holding and locking of one part in the other as you have in the hinge and the ball-and-socket of the mechanic. In every case the bones are held together, not by their own shape, but by ligaments and muscles. Consequently, any one of the bones may be dislocated from those next it without breakage; and when the muscles and ligaments are cut

through, or have been destroyed by maceration,
all the bones, between which any movement was
possible during life, separate from one another.

Not only is this so, but in no instance are
the movements of joints *limited* simply by the
shape of the bones—that is to say, they are never
brought to a stop by a part of one bone coming
into contact with the edge of another. Such a
contact would have caused a *sudden* check; and
this would have been attended with more or less
jar and with some danger of chipping and breaking the articular edges. The range of movement
of a joint is always regulated by the ligaments
or the muscles, not, directly, by the bones; and
the restraint thus imposed upon the movements
is brought to bear, not suddenly, but *gradually;*
somewhat like the effect of the "break" upon a
railway-train; while the cartilages between the
bones may be compared with the "buffers" between the carriages.

It is chiefly by means of strong LIGAMENTS,
or sinewy bands, passing from bone to bone, that
the shape of the plantar arch is maintained and
the movements of the bones upon one another
are regulated and limited. These ligaments are
numerous; but I will mention only two.

One, the *Plantar Ligament* (A, fig. 7), of
great strength, passes from the under surface of

the heel-bone, near its extremity, forwards, to the
ends of the metatarsal bones; in other words, it
extends between the lowest points of the two
pillars of the arch, girding, or holding, them in
their places, and preventing their being thrust
asunder when pressure is made upon the key-
bone (D); just as the "tie-beam" of a roof resists

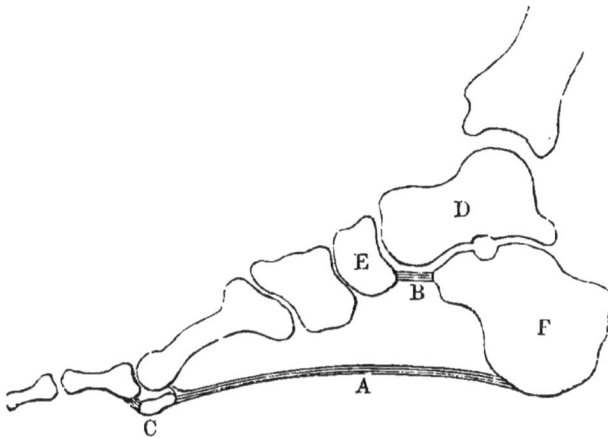

Fig. 7.

the tendency to outward yielding of the sides
when weight is laid upon the summit. The ligament, however, has an advantage which no tie-
beam can ever possess; inasmuch as a quantity
of muscular fibres are attached along the hinder
part of its upper surface. These instantly respond
to any demand that is made upon them, being
thrown into contraction directly the foot touches

the ground; and the force of their contraction is proportionate to the degree of pressure which is made upon the foot. Thus they add a living, self-acting, self-regulating power to the passive resistance of the ligament. In addition to its office of binding the bones in their places, the ligament serves the further purpose of protecting from pressure the tender structures—the blood-vessels, nerves and muscles—that lie above it, in the hollow of the foot, under the shelter of the plantar arch.

Another very strong ligament (B in the wood-cut) passes from the under and fore part of the heel-bone (F) to the under part of the scaphoid bone (E). It underlies and supports the round head of the astragalus, and has to bear a great deal of the weight which is transmitted to that bone from the leg. It does not derive the same assistance from a close connexion with muscular fibres as the ligament just described; but it possesses a quality, which that and most other ligaments do not have, viz. elasticity. This is very important, for it allows the head of the key-bone (D) to descend a little, when pressure is made upon it, and forces it up again when the pressure is removed, and so gives very material assistance to the other provisions for preventing jars and for giving ease and elasticity to the step.

A glance at the drawing will show you that here is a weak point in the foot. The head of the key-bone receives great weight from the leg, but is comparatively unsupported; and there is a considerable strain upon this part when the heel is being raised in walking. Moreover, a good deal of movement takes place between the key-bone (D) and the scaphoid bone (E), more than between any other two bones of the instep; and freedom in the range of movement is generally attended with some sacrifice of strength. The strong elastic ligament comes in therefore with peculiar advantage at this point; and it is underlaid, and additional support is afforded exactly when it is most required, by the tendon (b in fig. 12) of a strong muscle, the especial office of which is to assist in raising the heel and bending the instep, and which runs, from the back of the leg, behind the inner ankle, to the scaphoid bone.

Weak Ankle and Flat-foot.

In spite, however, of the thick elastic ligament and the strong tendon just mentioned, the joint between the astragalus or key-bone and the scaphoid bone still remains a weak point. The head of the key-bone, from being insufficiently supported or from being overweighted, is very apt

to descend a little below its proper level; the consequence of which is that the plantar arch is lower-
ed and the foot is flattened; and the more the foot
is flattened the weaker it necessarily is, because
the position of the bones then becomes less and
less favourable for bearing weight, and an increasing strain is thus incurred by the ligaments and
muscles. Hence the foot and ankle feel weak;
and the weakness is especially felt when the person endeavours to raise the heel, so as to mount
upon the balls of the toes, in walking. For the
performance of that movement with ease and
steadiness a well-formed plantar arch is essential;
and the person, whose feet are defective in the
manner we are considering, can never walk with
a bold, firm step. The movement in him may be
better described as a shuffling from one foot on to
the other, than as a walk. To this I will recur
again when I come to speak more of walking. The
defect, when slight in degree, is commonly called
"weak-ankle;" when more decided it is called
"flat-foot," because the sole is then nearly, or
quite, flat. The head of the key-bone, under
such circumstances, may even bulge downwards
and inwards, and form a prominence on the inner
side of the sole, so as to give more or less con-
vexity to the line on the inner side of the foot,
which should be concave.

The representation of "flat-foot" here shown was drawn from the foot of a labouring man in

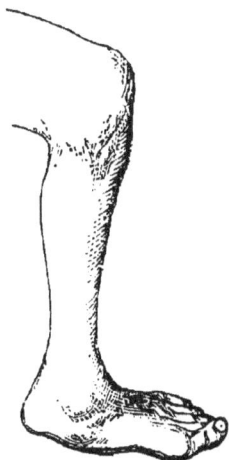

Fig. 8.   Flat-foot.

this county.   He said he believed the deformity was due to his having worn thick tight shoes when he was a growing boy.   He is most likely right in his opinion; for tight or ill-fitting shoes, cramping the feet and preventing the proper growth of the bones and the free play of the muscles, are a common cause of this evil.   This is so especially among the agricultural class, whose feet are, from an early period, enclosed in stiff unyielding leather cases that are enough to mar nature's best efforts to construct a plantar arch.

The same drawing shows that flat-foot is not the only deformity for which "high-lows" are answerable. Besides the almost total want of calf, which is due to the wearer being obliged to hobble along, whole-footed, with short feeble steps, it will be seen that the great toe has not been allowed to assume its natural straight line, but has been squeezed athwart the other toes, so as to lie almost at a right angle with the foot. No room at all is thus given for the second toe; it has been driven quite out of the field, and has been obliged to hide itself by bending down under the other toes. This is no uncommon state of things. Frequently it is attended with the formation of a painful bunion upon the prominent inner side of the ball of the great toe; and, in addition, there is sometimes a corn upon the first joint of the second toe, which is a source of so much inconvenience that I have known many sufferers glad to get relief by parting with the toe.

I wish I could hope that the days of highlows are numbered, and could believe that in the next generation they will be ranged with the things of the past, and that our children may know these enemies to the form of the rustic foot, only as objects to be gazed upon with feelings of astonishment and pity, just as we regard the perukes and the stays of our ancestors. There are, however, some practical difficulties in the way of the fulfilment of this charitable wish.

There are two periods of life at which FLAT-FOOT is most likely to be engendered. *First,* in infancy, if the child be put upon its feet too early, before the bones and ligaments are strong enough to bear the weight of the body. Therefore mothers should not indulge their anxiety to see their infants walk very early ; the pride attendant on premature success is liable to be followed by regret at finding that the children never walk well. Parents and nurses should be content to let the children crawl and roll about upon the floor, and should not encourage them to stand upright, especially if they be rather heavy or weak children. Children are quite sure to acquire the faculty of walking as soon as they are well fit to exercise it.

The *second* period is at about fourteen. The body attains a considerable increase of weight at this time, in consequence of the quick growth that takes place. We often remark that lads and girls of this age shoot up apace ; and their greater weight is not always attended with a proportionate·acquisition of strength. They are apt to be rather weak and ungainly in their movements ; and the weakness often shows itself in the foot, by a yielding of the plantar arch. Moreover, many boys and girls are, at this age, turned out into the world to earn a livelihood, and are obliged to be a good deal upon their feet, and perhaps, in addition, have to carry weights. Thus errand-boys, butchers' and bakers' boys, and young nursery-maids, are frequent sufferers in this way. The constrained positions in dancing, also, if enforced too much, or continued too long, so as to tire the feet, sometimes lead to the same result. On the other hand, moderate exercise of this kind is calculated to strengthen the foot and also the whole frame, and contributes much to improve the carriage.

This is not the place to enter into particulars of *treatment*. I will, therefore, merely remark that the common notion of supporting and strengthening the ankles by tight-laced boots is altogether a mistake, and must be ranked among the most influential of the causes which combine to spoil so many feet. It has its parallel in the idea of strengthening the waist by stays. The notion is, in both instances, fortified by the fact that those persons who have been accustomed to the pressure, either upon the ankle or the waist, feel a want of it when it is removed, and are uncomfortable without it. They forget, or are unconscious, that the feeling of the want has been engendered by the appliance, and that had they never resorted to the latter they would never have experienced the former; just as dram-drinking induces a recurrence to the stimulus by causing a sense of sinking when it is discontinued; and, for the same reason, the opium-eater can hardly exist without his drug.

The Movements of the Foot.

We come now to the MOVEMENTS of the foot upon the leg; and rarely do we contemplate anything more calculated to excite our admiration. Consider their variety, the rapidity with which they take place, in order to effect the requisite succession of positions in walking and running, and to adapt the sole to the inequalities of the surface on which we tread; and remember the great weight which has to be sustained while these movements are going on: yet, how seldom is there a failure.

This combination of variety of movement with security is effected by the employment of *three* joints, each of which plays in a direction different from the others, while all act harmoniously together.

One of the three joints—strictly called the "ankle-joint"—is between the leg-bones and the foot-bones, that is, between the tibia and fibula, above, and the astragalus beneath. By means of it the foot may be bent or straightened upon the leg; in other words, the toes may be raised or depressed. In this movement the heel participates, being depressed when the toes are raised, and *vice versâ*. A *second* joint is between the astragalus and the heel-bone. It permits the foot to be rolled inwards or outwards upon an antero-posterior axis; so that the sole may be turned inwards, with its inner edge upwards, or may be turned down so as to be placed flat upon the ground. A *third* joint is between the first and second row of tarsal bones—that is, between the astragalus and the heel bone, behind, and the scaphoid and cuboid bones in front. It permits the degree of flexure of the tarsal or plantar arch to be increased or diminished.

Had the several movements which are requisite for easy walking all taken place in one joint, that joint must necessarily have been very insecure; indeed, it must have been a "ball-and-socket" joint, and we should have been poised upon our feet in the state of what is called "unstable equilibrium"—a state quite incompatible with security or strength, and which would have rendered the assistance of the upper limbs essential to either standing or walking.

An instance of a similar kind of mechanism to this of the joints between the foot and the leg is presented by the mode in which the head is secured upon the back-bone. We can nod the head upwards and downwards; we can turn it to either side in so free a manner that we are able to command with our eyes the whole circle in which we sit simply by the movements of the head; and we can incline the head to the right or to the left. Any of these movements may be made very quickly; and there is a separate joint or joints for each of them. Thus, the *nodding* movement takes place between the head and the first vertebra or uppermost bone of the spine; the *turning* of the head from side to side takes place between the first and second vertebræ, the head with the first vertebra rotating upon a pivot projected upwards from the second vertebra; and the *inclination* of the head from side to side takes place by movements of the second vertebra upon the third, of the third upon the fourth, and so on. The result is that, although the movements are thus varied, they are free as well as rapid. Yet the head is so well poised and so strongly fixed that the neck is able to bear it all day long without fatigue; and, as though the weight of the head, which is by no means inconsiderable, were not enough for the neck, we are in the habit of

selecting this as the part upon which to carry bur-
dens. One never feels so strongly impressed with
the carrying capabilities of the neck and the ankle,
as when following men and women in mountain
districts toiling up and down the hills under great
bundles of hay, baskets full of bitter beer, and
various things intended to minister to the comfort
and luxury of travellers and the inhabitants at
the top. So effectual, indeed, are the provisions
for security that, notwithstanding the freedom and
variety of their movements, the joints of the foot
with the leg, and of the head with the spine, are,
in proportion to their size, the strongest in the
body.

I have stated the movements that take place
in the three joints of the foot with the leg in a
simple manner, for the sake of avoiding confusion.
In reality, however, they are not so simple, but
very difficult to analyse and make out correctly.
The difficulty is due, partly, to the close proximity
of the joints to one another, which renders it no
easy matter to distinguish the movements of one
from those of the others, and, partly, to the fact
that the movements in each joint are a little oblique.

In the latter respect the foot-joints resemble
most of the others in the body; and it is this *obliquity* in the movements of the joints, added to

the *curves* and *twists* in the shape of the bones,
that constitutes one of the chief difficulties in
investigating and clearly understanding the mechanism of the human frame. It has been said
that "Nature abhors a vacuum:" it may with
equal truth be said that she abhors a straight line.
In the Human Skeleton, at any rate, all the bones
are bent and twisted, some in two or three directions; and the surfaces by which any bone is
jointed to the adjacent bones, are invariably oblique with regard to each other.

Take, for instance, the *tibia*, or large bone of
the leg, of which a front view and an inner side

Fig. 9.

view are given in the drawings. The tibia is a

column transmitting weight from the thigh to the
foot; and in any machine of man's construction a
column fulfilling similar purposes would be made
straight and of uniform diameter throughout. The
bone, on the contrary, does not present the same
thickness at any two parts of its length.   It has
a distinct bend, forwards, in nearly its whole
length (fig. 10): there are lateral curves, alter-
nating like those in the letter S, seen along its
front (fig. 9): and the articular surface at the

Fig. 10.

lower end is placed obliquely with regard to that
at its upper end, in consequence of a twist in
the shaft, in such a manner that when the
hinder surface of the upper end of the bone rests
upon a board, the lower end touches the board
only by its outer corner (fig. 10). This disposition of the lower end, I may remark, assists to
give the foot a slant outwards from the heel to
the toe, so that when we stand, with the heels
together, the great toes of the two feet diverge
a little from one another.

Moreover, the surfaces by which the tibia is jointed with the thigh-bone at the knee are ar- . ranged with a varying degree of obliquity, so that the relation of the leg to the thigh varies somewhat in different positions of the limb. For instance, when we stand upright, the *thigh* slants *in*wards from the pelvis, and the *leg* descends in a *vertical* direction to the ground. While, however, the knee is being bent the leg is carried, not in a vertical plane, but a little obliquely, so that the lower part soon begins to slant *out*wards; and when the knee is fully bent the obliquity of the leg and that of the thigh correspond, and the leg is, as it were, folded up against the thigh. The heel is thus brought up, not to the middle line of the body, but to the hip, and we are enabled to sit with the hips upon the heels, as

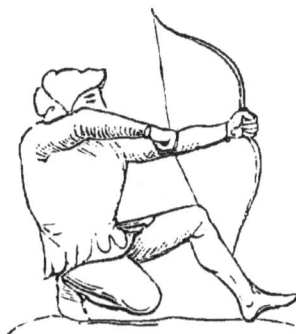

Fig. 11.

the Japanese are represented doing, or with one

hip upon one heel—a position in which our riflemen are trained to take aim, and in which their predecessors with the arrow were wont to shoot, as is shown by the accompanying sketch of a bowman (fig. 11), taken from one of the Ægi- netan marbles in the Glyptothek at Munich.

A variety of purposes is attained by the cur- vilinear shape of the bones and the obliquity in the movements of the joints. Not the least of these is the appearance of elegance and ease which is given to the whole frame, both when it is at rest and when it is in motion. In order that you may fully appreciate this result, I would ask you, the next time you are in a gallery of antique sta- tuary, to contrast the figures which the Egyptians have left us with those by the Greeks. In the former you will find that straight lines and right angles prevail : the figure sits, probably, bolt upright, with the elbows, hips, knees, and ankles bent at right angles: the fingers commonly run straight forwards; and a hand is often laid upon each knee, the limbs of the two sides being quite symmetrically placed. Such statues may be im- posing ; but they are stiff and unnatural. They represent positions which the body rarely assumes; and they, certainly, are far from pleasing. Very different is the Greek statuary. A correct re- presentation of nature is the great difficulty and the highest consummation of art; and the Greeks evinced their greatness in art by a true appreciation and close imitation of natural form. The position of their figures is life-like; and, therefore, we love to contemplate them. The outline in them exhibits a graceful disposition of curves and obliques; and it is because the great sculptors of Greece were, in this and in other respects, so true to nature that their works have commanded the admiration, and served as models for the imitation, of all succeeding ages.

It is one of the master results of creation, and one of the peculiar marks of creative genius, that *perfection* and *beauty* are usually presented together. As truth is the soul of eloquence, so is perfection the soul of beauty. The works of nature are beautiful because there is so much excellence in them, such admirable adaptation to their purpose; and we find the works of man beautiful only so far as they are correct imitations of their great originals in nature, or show some approach to nature's excellence. And man is the most beautiful object in nature because he is the most perfect, that is, because the purpose of his existence is the highest, and because his physique exhibits the most marvellous moulding to adapt it to its high purpose; because, in short, in him the material is wrought to such a

point of refinement as to be the receptacle and
minister of the immaterial.

The movements of the three joints between
the foot and the leg take place in harmony.
The following is the order observed. The raising of the *heel* is accompanied by a rolling of
the foot *in*wards, and by an increased *flexure* of
the plantar arch ; and the raising of the *toes* is
accompanied by a rolling of the foot *out*wards
and a *straightening* of the sole.

The Muscles of the Leg and Foot.

The *first* series of the movements just described is effected, mainly, by three muscles.   Of

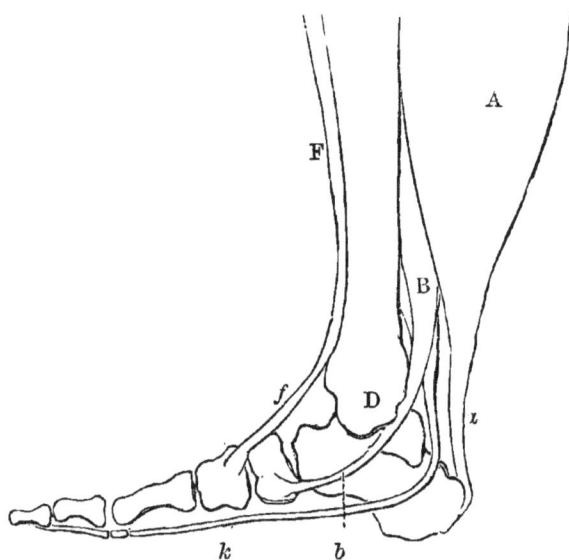

Fig. 12.

these one (A, fig. 12) raises the heel while the other two (B, fig. 12, and C, fig. 13) raise and support the ankle. The muscle which acts upon the heel is one of the largest and most powerful in the body; and well it may be, for in raising the heel it has to raise the whole weight of the body. Its fibres, accumulated at the middle and upper part of the leg, form the "calf;" below they taper into a thick tendon (a) connected with the hinder extremity of the heel-bone, and called the *Tendo Achillis*. The name, it need scarcely be said, refers to the tale of Thetis holding her son Achilles by this part when she dipped him in the river Styx. Her hand prevented the part from coming in contact with the water; and so it did not partake of the invulnerability which was conferred upon the rest of his body by the immersion. We read, accordingly, he was finally killed by a wound in the heel[1].

[1] It does not appear that the legend is based upon any peculiar ideas of susceptibility attached to the heel among Eastern nations; nor can the passages in Scripture, that the Serpent shall bruise man's heel (Genesis iii. 15); "For the greatness of thine iniquity are thy heels made bare" (Jeremiah xiii. 22), be adduced as indicating the existence of such an idea. There are some other myths resembling this one of Achilles; but in them a different part of the body missed the protecting influence. Thus, Ajax was wrapped by Hercules in the skin of the Nemæan lion, and was, thereby, rendered invulnerable, except at the pit

The other two muscles (B and C) also descend from the leg and terminate in tendons (b and c) which pass, one on either side, behind the projections (D and E) which we call respectively the inner and outer ankle, to the inner and outer edges of the instep. They assist to raise the ankle, and support it so as to prevent its swerving from side to side; and they permit it to play to and fro upon them, like a pulley upon ropes running under it, in a safe and easy manner. The inner (b, fig. 12) of the two tendons passes, as before mentioned, beneath the head of the key-bone, and adds greatly to the strength of the arch. It is, moreover, the chief agent in effecting the two movements which are associated with the elevation of the heel, viz. the turning of the sole inward and the flexion of the foot.

The *second* series of movements—the raising

of the stomach where the edges of the skin did not quite meet; and he killed himself by running his sword in there. In the *Niebelungenlied*, the hero, Siegfried, is represented to have rendered himself invulnerable by smearing himself with the blood of a dragon which he had killed. A leaf, however, adhering to his back, prevented the contact of the fluid with one spot. The secret was unwarily communicated by his wife Krimhild to his enemy Hagan, who took advantage of the information to plunge his sword into the fatal spot while Siegfried was stooping down to drink at a rivulet.

The lesson inculcated by these myths seems to be that all men, even heroes, have their weak points.

the toes, the turning the sole downwards, and the straightening the foot—are effected by two mus-

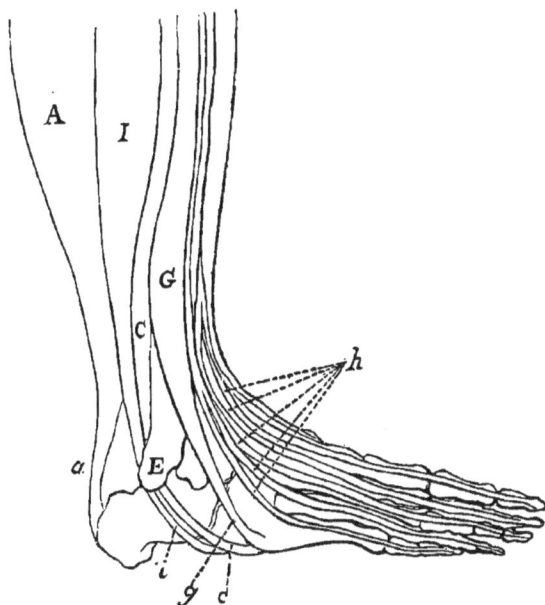

Fig. 13.

cles (F, fig. 12, and G, fig. 13), the tendons (f and g) of which pass, one in front of the inner ankle, and the other in front of the outer ankle, to the respective edges of the instep. These require much less power than their opponents; and the muscles on the front of the leg are, therefore, smaller and weaker than those behind.

A question of practical interest here suggests itself. How is the balance between these antagonistic muscles maintained, and the proper position of the foot preserved? If the muscles which cause the elevation of the heel and the other movements associated with it are so much stronger than those which produce the opposite series of movements, and if, as we know to be the case, muscles are always, even when a limb is at rest, contracting with a certain amount of force, why do not those of superior power gain and maintain the ascendancy, and hold the limb in the position to which they have a tendency to draw it? And why, in this instance, are not the feet kept with the heels raised and the soles inturned and bent? The reply is, that the ill consequence suggested is prevented, and a proper adjustment between the opponent sets of muscles, in this and other parts of the body, is effected through the medium of the nervous system. That system institutes friendly relations, and compels an orderly and harmonious action of the several muscles; and it does so by frequently exerting its influence upon them, keeping them in drill, as it were, and enforcing the habit of yielding in a kindly manner to one another.

You have often observed, and perhaps wondered at, the almost incessant, semi-involuntary and, seemingly, meaningless movements of infants, especially the peculiar sprawlings out of their fingers and toes. Now these are for the purpose

of keeping the different sets of muscles in practice
and in order, till the will acquires a due control,
when they gradually cease. They are going on
before birth as well as afterwards; and when they
are deficient, or when they take place irregularly,
in consequence of an imperfection in the nervous
system, the limbs are liable to become deformed.
The feet, under these circumstances, are often
drawn into the very position I have just mentioned; the sole is turned inwards and upwards,
so as never to touch the ground; the heel and
the toes are approximated; and the foot rests upon
the ground on the outer side, or quite on the fore
part, of the instep. Such a condition constitutes
one of the most common forms of what is called
" club foot." Children are often born with one or

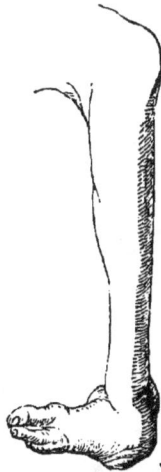

Fig. 14. Club-foot.

both of their feet thus distorted. Happily, however, if they be submitted in time to the modern improved modes of treatment they may usually be set right. The accompanying woodcut gives a sketch of the foot of a young woman who had not the good fortune to be thus attended to.

The muscles compose the flesh or chief part of the bulk of a limb. The " calf" is almost entirely made up of the fibres of the " calf-muscle." But at the ankle there are no muscles. As they descend the leg, all the *muscular* fibres disappear, and there are only *tendons*. These, though much thinner than the muscles, are very strong; and they are the cords or ropes by which the muscles pull upon distant parts. As they pass over the ankle they are strapped down close to the bones by means of stout sinewy cross-bands, which prevent their starting from their places when the muscular portions pull at them.

Two especial advantages result from this arrangement.

First, the lower part of the leg and the ankle are reduced in size. Thereby the resistance to the passage of the limb through the air is lessened; and when it is upon the ground, the leg is less in the way of the other foot which is swinging, to and fro, beside it. An elegance of shape is also thereby imparted. The "pretty ankle" owes much of its charm to the mode in which the tendons are disposed. How comparatively thick and clumsy would the ankle be if the tendons of the toes took the straight course represented by the line a in the drawing, instead of being bound down, as they are, to the curve of the ankle!

Secondly, the obliquity with which the tendons run to their insertions is increased by this arrangement; and the velocity of the movements to which they minister is increased also. True, a loss of strength is involved in such a disposition, but the gain in velocity is of more importance.

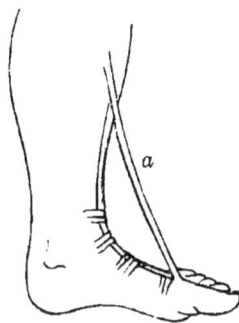

Fig. 15.

If (to refer again to the diagram, fig. 15) the tendon ran in a straight course from the front of the leg to the great toe, the angle at which it joined the toe would enable it to act with more strength; but the movements connected with it could not be so quick as they now are.

H.   4

We find in the construction of the human frame many instances in which strength is sacrificed to rapidity of movement in this and other ways. Scarcely any conceivable amount of strength, for instance, would be an adequate compensation for a loss of that celerity of movement of the hand which enables us to strike a blow and drive a nail. No wonder, therefore, that strength is here sacrificed to obtain celerity. And the same principle holds good for other parts.

The length and direction of the heel affords a good illustration of the principle of which I am speaking. When the heel-bone runs out to a considerable distance, and nearly straight, behind the ankle, as it does in some of the lower animals and in the inferior races of mankind, it presents a better leverage to the calf-muscle, which is, then, enabled to raise the ankle with a less amount of effort; but there is proportionately less velocity. Accordingly, in the more perfectly formed foot, such as we find it in the higher races of mankind, the heel-bone, instead of running out backwards, descends very obliquely, almost vertically.

In this instance, the loss of strength, which is thus incurred for the purpose of acquiring celerity in movement, is usually compensated for by the greater development of the calf-muscle. Hence the high heel and the well-developed calf go together; and, like most of the other features of good bodily formation, they are, on the whole, best marked in the nations which are endued with the highest intelligence, and which are, in this way, physically, as well as mentally, qualified to occupy the foremost places in the human family. Thus, we may mark a relation between the heel and the brain; and, as the comparative anatomist is able by the inspection of a bone to trace out the skeleton to which it belonged, so might it be possible for the human anatomist, by observing minutely the peculiarities of the heel and the other features of the foot in any particular race of men, to form some estimate of the capacity and conformation of the skull, and thereby, of the amount of intelligence.

Contrast the foot and leg of the EUROPEAN (fig. 16), as represented in the drawing reduced from the Farnese Hercules, with those of the NEGRO (fig. 17), the drawing of which was taken from a native of Sierra Leone. In the former the leg is plump and the calf well developed; the foot is compact and well arched; the heel descends nearly vertically; and the inner ankle stands clearly out and is raised high above the ground. In the Negro the leg is thinner and the calf is not so well defined; the foot is long, flat, and sprawling; the heel is more horizontal;

and the inner ankle does not show clearly, and
almost touches the ground.

Fig. 16.   Fig. 17.

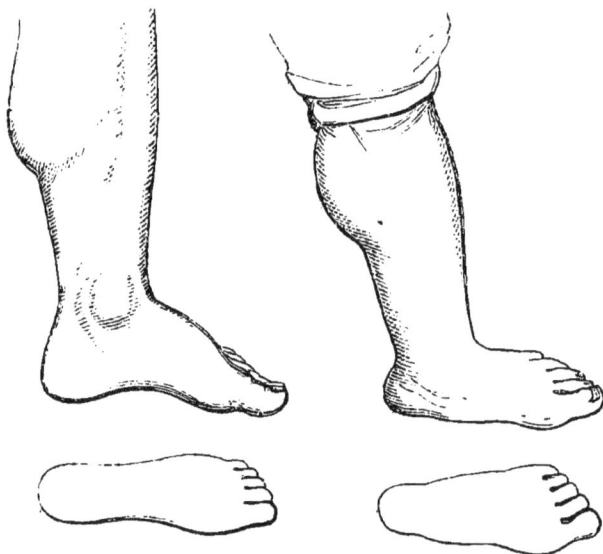

Fig. 18.   Fig. 19.
European.   Negro.

Contrast also the outline (fig. 19) of the foot of
the same Negro with that (fig. 18) of an Englishman. Both were traced upon the ground, and reduced upon the same scale. The Negro was 5 ft.
2 in. in height; the Englishman was 6 ft.; both were
of the same age: yet the Negro's foot was considerably the larger. It was 11 inches long, $3\frac{1}{2}$ inches
across the middle of the instep, and $10\frac{1}{2}$ inches
round the balls of the toes. Whereas the English-
man's foot was less than $10\frac{1}{2}$ inches long, was $2\frac{1}{2}$

inches across the middle of the instep, and $9\frac{1}{2}$ inches round the balls of the toes. Even in this simple outline how much less shapely is the African's foot. Some allowance must be made for the fact that the Negro was more accustomed to go barefooted than the Englishman; and the pressure of the boot or shoe has, in some degree, the effect of giving compactness to the foot.

In the native AUSTRALIAN the leg is commonly still more lanky, there being less calf than in the African; and in the MONKEY the heel is quite horizontal, the sole is flat, and the muscular fibres of the leg are continued low down, close to the ankle, instead of being concentrated higher up ; so that the leg has nearly the same thickness from the knee to the foot, and there is no calf at all. Indeed, in the GORILLA (see fig. at page 90) the circumference of the leg increases towards the ankle. Thus, the calf may be regarded as the characteristic of MAN ; and a well-developed calf is a characteristic of the higher members of the human species. The pride, therefore, which is felt in a well-formed leg is not altogether a senseless folly, but finds some excuse in the fact that its foundation lies deep in the laws of physiology and ethnology. It must be confessed, that the fashion which, in the last century, dictated the knee-breeches, the silk stocking, and the shoe, evinced a truer appreciation of the dignity and beauty of the human figure than do the modern investments, which quite cover up the limbs, encumbering their movements and hiding the beauty of the leg and ankle.

In the addition of the *high heel* to the shoe we recognise an effort to improve upon the original, by exaggerating one of the peculiar features of the human foot; but it results in a failure, as is invariably the case with such strainings after a greater perfection than nature has given. It increases the apparent height of the person and of the arch of the instep; but it throws the weight too forward upon the toes, and detracts from the length and security of the step. Moreover, by causing disuse of the elevators of the heel, it interferes with the full growth of the calf.

This is, however, a harmless piece of vanity in comparison with the monstrous efforts of the

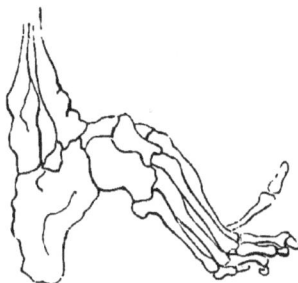

Fig. 20.   Chinese.

Chinese to mould the foot to their ideal by squeezing the heel and the toes together. They effect this to such a degree that (fig. 20) the heel-bone descends vertically from the ankle, the plantar arch is bent to an acute angle, and the foot is so crumpled up that all movement in it is effectually prevented, and the part is reduced almost to a mere stump. These observant and ingenious people have caught, it may be, the idea that compactness, elevation of instep, and sudden descent of heel are characteristics of the well-formed foot, and may urge that they are helping nature to perfection in the direction which she has herself indicated. But in their silly attempt at the preternatural, in this impious use, as it were, of fire stolen from heaven, they simply burn and cripple themselves, and render themselves ridiculous, and give to all other nations the much needed lesson that it is enough for man to follow as a humble imitator of his Maker's works, and that his attempts to alter, or improve upon, any part of the wondrous design of creation will assuredly have the effect of spoiling and defacing it[1].

[1] It is a remarkable statement by a correspondent in *The Times*, Jan. 14th, 1861, that in the pillage of the Summer Palace of the Emperor of Pekin "all the ladies of the Court must have had natural-sized feet, all the slippers found in their rooms being large ; not a single cramped-footed shoe was seen."

It seems that the several races of mankind are usually rather proud of their peculiarities, and that each has an inclination to make much of, and artificially exaggerate, the points in which it differs from the others. Thus the Chinese are remarkable for the spareness of their hair and the smallness of their feet; so the men shave their heads, leaving only the pig-tail, and the women squeeze up their feet in the remorseless manner we have seen. The Singhalese, who are flat-footed, are said to consider it one of the requisites for a 'belle' that the soles of her feet should not have any hollow. The red Indians of America delight in staining and painting their skins of a lively red colour. The Columbian tribe of Indians increase the natural lowness of their forehead by flattening it out in infancy, and succeed in bringing about a deformation of the skull almost as remarkable in its way as is the effect of Chinese cramping upon the foot. These people also take pains to reduce the small quantity of hair upon their eyebrows, lips, and chin, by plucking it out.

I will briefly draw your attention to one other point in the anatomy of the foot; and that is, the mode in which the "metatarsal" bones are jointed with the "tarsal." If you take hold of the ends of the metatarsal bones—in other words the "balls"—of the great toe and of the two toes next to it, in your own foot, you will find that you can move them scarcely at all; they are firmly set upon the rest of the foot, almost as though they formed one piece with it. If you then try the end of the metatarsal bone of the fourth toe you will be able to move it a little upwards and downwards; and in the case of the little toe the movement is still more distinct. This difference depends upon the mode of construction of the joints of the metatarsal bones with the tarsal, which is easily understood by the aid of the accompanying drawings, representing sections, from above downwards, through these joints. In 21, 22, and 23, which are the tarso-metatarsal joints of the great toe and the two next it, the opposed surfaces of the bones between *c, c,* are quite flat, so that the only movement that can take place is a slight sliding of one bone upon the other, just enough to assist

in breaking the jar, but not enough to interfere
with the firm basis of support which these toes

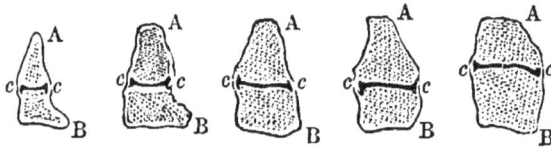

Figs. 25.   24.   23.   22.   21.

are required to afford to the plantar arch in consequence of the great stress of the weight in walking
being borne upon this side of the foot. In No. 24,
which is the joint of the ring toe, and still more in
No. 25, which is the joint of the little toe, the end
of the metatarsal bone (A) is rounded and is received
into a corresponding concavity or cup in the tarsal
bone (B). This allows a slight revolving of one bone
upon the other to take place, and permits the
movement which you discover when you grasp the
balls of these two toes between your fingers. The
outer part of the foot needs not to be so strong and
firm as the inner part, because it does not lie so
nearly in the plane of gravity during walking; and
the provision just described, which permits some
movement in the outer two metatarsals, enables
the balls of the toes to adapt themselves to inequalities on the ground, and to share more equally,
under various circumstances, the weight which is
thrown upon them.

Standing and Stooping.

When we STAND straight upright (fig. 26) the centre of gravity of the head is directly over a

| Figs. 26. | 27. | 28. | 29. |
| Standing. | Bowing. | Stooping. | Squatting. |

point midway between the two ankles; and the plane of gravity, represented by the vertical line in the figure, descends, from the head, through the spine, pelvis, and lower limbs, to the key-bone of the instep. And you observe that, between the head and the ankle, the skeleton is not quite straight, but is arranged in six curves, which are, alternately, in front of and behind the line of gravity. Of these curves the upper three are in the spine. They are well marked; the uppermost (*a*) is in the neck and is directed forwards; the next (*b*) is in the back and is directed backwards; the third (*c*) is in the loins and is directed forwards. The fourth curve (*d*), less distinct than those above it, is in the pelvis and is directed backwards. The fifth and sixth curves are very slight; the fifth (*e*), directed forwards, is at the hip-joint; and the sixth (*f*), directed backwards, is at the knee. The last two curves, though slight, are not unimportant; and they contribute very much to our comfort and to prevent fatigue when we are standing: they do so in the following way. The strong ligaments of the hip are placed towards the *fore* part of the joint, that is, in *front* of the line of gravity; and the strong ligaments of the knee are placed towards the *back* part of the joint, that is, *behind* the line of gravity. It follows that when these joints are fully extended they are "locked," as it is termed, just as is a hinge when opened to a little beyond the straight line; and, by this means, the muscles are set at rest, and we are able to maintain the erect posture, for some time, steadily and without fatigue.

When standing upright in this way, at rest on both legs, or on one leg in the military position of "at ease," and the muscles are off their guard, if a sudden and unexpected, though slight, pressure be made upon the ham, so as to bend the knee a little and throw the joint in front of the line of gravity, the man will drop, unless the muscles come quickly to the rescue—a tendency which has not escaped the observation of school-boys.

In BENDING or BOWING (fig. 27) the head is carried forwards; and, to maintain the balance, the opposite pole of the trunk is carried backwards, so as to preserve the line of gravity still over the ankles.

In STOOPING (fig. 28) or SQUATTING (fig. 29), as in picking up any thing from the ground, the lower limbs and the trunk are bent in a zigzag manner; the heels are raised; and the plane of gravity falls, in front of the ankles, over the balls of the toes. Now we recognise one of the advantages which accrues to man from the great length of his thigh. For the head and upper part of the trunk are advanced so far in *front* of the feet, that it would be impossible to maintain a balance at all, even upon the balls of the toes, and we should necessarily fall forwards, were it not that, owing to the length of the thigh, the lower part of the trunk is carried backwards to a plane *behind* the heels, and so serves to maintain the equilibrium.

Walking.

Let us next consider the part which the foot performs in WALKING. To understand this it is necessary to consider its positions and movements in the several stages of a step. When first placed

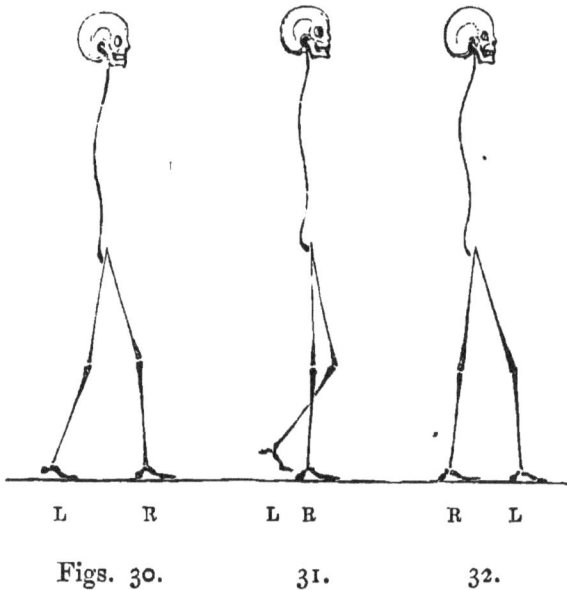

L   R   L   R   R   L

Figs. 30.   31.   32.

Walking.

upon the ground the foot (R, fig. 30) is a little in advance of the body; and the heel comes first (fig. 33) into contact with the ground. The toes quickly follow; and the body, then, passes, vertically, over the ankle and the key-bone of the instep. The foot

(R, fig. 31 and fig. 34) now rests steadily upon the heel and the balls of the toes; the other foot (L) leaves the ground, so that the whole weight is borne by one foot; and the plantar arch of that foot expands a little, so as to cause slight lengthening of the foot, under the weight that is laid upon it. Much yielding of the arch is, however, prevented by the ligaments that brace the arch (fig. 7), and by the muscles that are disposed be-

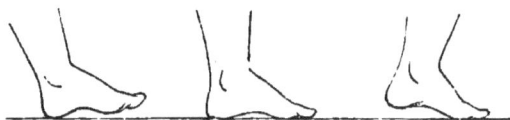

Figs. 33.   34.   35.

Walking.

neath it. Next, the heel (fig. 35) is raised by the action of the calf muscle, and the weight of the body is thrown forwards, over the balls of the toes, while the other foot (L, fig. 32) is carried onwards, and is placed upon the ground ready to receive the weight and commence its carrying work. When this has been done the foot is withdrawn from the ground; and, in the withdrawal, a final impulse onward is given, so as to throw the weight of the body fairly over to the other foot. The fore part of the foot is then raised, and the knee is bent a little. By these means the toes are kept clear of the ground, while the foot is swung forward, beside the other, so as to be ready again to rest upon the ground and bear the weight of the body.

In each complete step, therefore, there is a period during which the foot rests upon the ground, and a period in which it is swinging in the air. In walking the former period is considerably longer than the latter; and at the commencement, and at the end, of that period (figs. 30 and 32) the other foot is also upon the ground, so that it is only during the middle of the time (fig. 31) in which the foot rests upon the ground that it has to bear the whole weight of the body.

Running.

In RUNNING the process is much the same as in walking. The chief difference is that, whereas in walking *both* feet are never *off* the ground at the same time, and both are *upon* the ground at the beginning and end of each step; in running *both* feet are never *on* the ground at the same time, and both are *off* the ground, and the body is flying unsupported through the air, at the beginning and end of each step (figs. 36 and 38). Thus, you may always distinguish running, though it be ever so slow, from walking, because, in the latter,

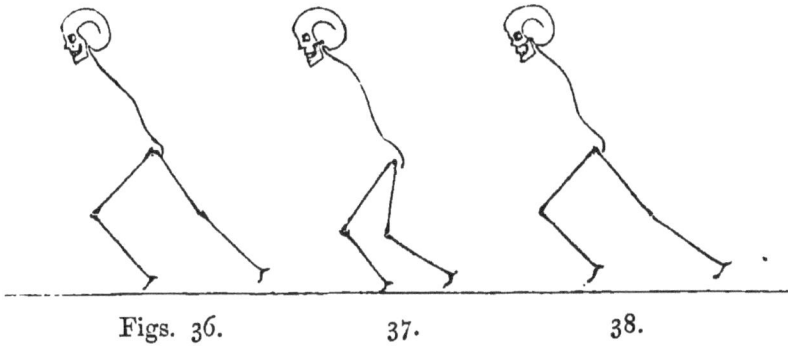

Figs. 36.   37.   38.

Running.

the two feet are upon the ground at the same time; while, in the former, only one foot touches the ground at a time.

The period during which the body rests upon the ground in running is comparatively very short, being merely the time when one foot is set down in the middle of each step (fig. 37); and great force has, consequently, to be exerted to propel the body through the air during the whole remainder of the step. Hence the exertion of running is much greater than that of walking. In slow running the same parts of the foot are applied upon the ground as in walking, and in the same order; but in quick running the balls of the toes only touch the ground. The quicker we run the shorter, relatively to the rest of the step, is the time during which the foot rests upon the ground, and the greater, consequently, is the effort.

After the foot leaves the ground, in running, it

H.   5

is thrown up behind; and, at the same time, the fore part of the sole and the toes are turned a little obliquely *in*wards, so as to prevent their catching against adjacent objects. If the toes were turned *out*, when thrown up behind, it would present a.very awkward appearance, and we should frequently be tripped up by their coming in contact with substances near which we pass. While the foot is being swung forwards the toes are gradually turned a little the other way. Thus, by the time they pass the other leg the toes have lost the inclination inwards, and are directed straight *for*wards; and when the foot has reached a point in advance of the other leg, and the sole is preparing to present itself to the ground, the toes are turned a little *out*wards. This turning of the foot *in*wards and *out*wards during its movement *back*wards and *for*wards, in each step, is a graceful movement, and may be compared to the "feathering" of an oar. It takes place, also, in walking, but is less marked than in running; and in many persons it can scarcely be discerned during walking.

The distinction between the paces of other animals resembles that between the walking and the running of man, and is equally definite. Take, for instance, the WALKING, TROTTING, and GALLOPING of the Horse. In WALKING the fore and the hind limbs of the *same* side are moved together, or nearly together, but they do not leave the ground till the limbs of the opposite side are placed upon it; so that at one period all four limbs are upon the ground together. In TROTTING the fore and the hind limbs of *opposite* sides move together; but, as in walking, neither of them are withdrawn from the ground till the opposite one has reached it[1].

In GALLOPING, or CANTERING, the horse springs or bounds with all four limbs at the same time; all the feet are thrown up nearly together; all are off the ground together; and all reach the ground again nearly at the same time ready for another spring. I say that the feet are all thrown up *nearly*, and not *quite*, together, because the fore and the hind limbs of one side take the precedence a little of the others, or "lead," as it is called. The trained horse is taught to lead, habitually, with one, usually the right, side, because the motion is more steady when the horse is accustomed to gallop in one way than if he be allowed to vary it. Directly the horse begins to gallop, the rider knows, by the motion, whether he is leading with

---

[1] In WALKING the hind leg moves first, then the fore leg of the same side; and both reach the ground before the hind leg of the opposite side is raised. So that at one time there are three feet on the ground, at another two, but never less than two.

In TROTTING, especially quick trotting, one foot is raised at the same instant that the opposite one is put down. This renders it difficult to make out the sequence of the movements.

the proper leg. In some animals, as the DEER, the two fore and the two hind feet move together exactly in galloping. Anthony Trollope tells us that in Panama, Cuba, and other Spanish countries in the West, the horses are "taught to pace, that is, move with the two off legs together, and then with the two near legs. The motion is exceedingly gentle, and well fitted for those hot climates, in which the rougher work of trotting would be almost too much for the energies of debilitated mankind." This *pacing* is probably only a quick walk.

When we walk the heels follow one another nearly in a straight line, as is shewn by "walking a chalk," or more readily by walking along the line between the curb and the flagstone pavement; and the plane of gravity of the body corresponds with this line. There ought, therefore, to be no perceptible *swerving* of the trunk from side to side in walking. There should, also, be scarcely any *rising* or *falling*; inasmuch as there are provisions in the mode of bending the limbs (which I cannot here discuss) to prevent the body from being moved up and down during the step. The head and shoulders should be carried along nearly in a straight line. If it were otherwise, if they were moved in a zigzag or undulating manner, from right to left, or up and down, the space traversed in a given distance would be much increased, and there would be a proportionately greater expenditure of muscular force. By a beautiful combination of movements this is prevented, and a rectilinear course is maintained, while the weight of the body is transferred from foot to foot, in a succession of steps.

Only observe a good walker for a minute or two, and you will see how straight the head is carried along; and when your eye falls upon a person who "rolls in his walk" you perceive how ungainly are his movements, and you have an instinctive feeling that he is an awkward fellow. Whether you are disposed to make an exception in favour of the British tar, in consequence of his many other good qualities, I must leave you to judge. His peculiar gait on shore is probably due to his not being sufficiently practised in straight walking to counteract the effect of the lounging manner and general disregard for appearances which he acquires on board ship. Whatever the reason may be, though he has the better of us in a storm at sea, he certainly does not always appear to advantage on *terra firma*. Now that a general improvement in gait and step may be expected among landsmen, as a result of the volunteer movement, it becomes still more desirable that the sailor should participate in the good influences of the drill.

Although the heels follow one another in a line the toes diverge a little from the line, because the foot slants, as I have just said, somewhat *out*wards when it is placed upon the ground. It results from this position of the foot that the weight of the body descends upon it with a slight obliquity, *in*wards as well as forwards; and that is precisely the direction in which the foot is best prepared to receive weight. For, when the foot rests upon the ground in this position all the ligaments on the inner side (and they are very strong) as well as those beneath, are on the stretch; and the joints, with the exception of the ankle-joint, are as it were locked, so as to afford a secure, steady basis of support to the leg. When the weight of the body descends upon the foot in the direction mentioned a sprain rarely occurs. It is when the weight falls in the opposite direction, that is, more or less obliquely *out*wards, and throws the ankle out, that a sprain easily happens. Thus a slight inequality of the ground, or any other cause that tilts up the inner edge of the foot, is likely to lead to a sprain, especially when we are going down hill or down steps.

Here let me remark that a SPRAIN is the result of a stretching of some ligament, or other part, caused by an undue force being brought to bear upon it. The ligaments are very strong,

and under ordinary circumstances are not very
sensitive; and they are capable of offering great
resistance to force applied in the direction in
which they are calculated to meet it. But, if the
force be applied in a direction in which they are
not calculated to meet it, they are easily injured,
and they become, then, very painful. The same
is, also, likely to occur if the force be severe or
sudden.

The muscles are a very great assistance to the
ligaments, forasmuch as, by placing and retaining
the joints in proper positions, they regulate the
direction in which forces are brought to bear upon
the ligaments. Moreover, by steadying or bracing
the joints, they check or break the force and
prevent its being suddenly imposed upon the ligaments. And the muscles, by virtue of their contractile property, have the capability of becoming
tight in any position of the joint, which is an im-
mense advantage; whereas a ligament having no
contractility and, usually, no elasticity, is tight
only in one position. The office of a ligament is
to limit the movement of a joint in a particular
direction; and, till the joint has assumed a certain position—till it is bent or straightened to a
certain angle—the ligament does not come into
play. During the bending or straightening of a
limb the muscles regulate the movement, and

bring it to a stop or check it before it has gone
to its full extent; and, thus, the ligament is re-
lieved from that sudden imposition of force which
would result if it were required to check the
movement of a joint in its full swing.

Accordingly, when the muscles are prepared
and in proper action, that is, when they place
the joint in a suitable position and duly support
or brace it, a sprain very rarely occurs. It is
when the muscles are unprepared, when we make
a false step, or when the foot encounters an unexpected obstacle, and the weight falls suddenly
upon the ligaments in an unfavourable direction,
that a sprain occurs. A man jumps from a considerable height, or descends deep steps with a
heavy weight upon his back, and no harm results; but he slips off the curb-stone, or treads
unwarily upon a piece of orange-peel, or turns his
foot hastily, to avoid some object on the ground,
and sprains his ankle.

In order that they may do their work well,
be alert, and maintain good guard, the muscles
need to be kept in practice. A person unaccus-
tomed to throw a ball is very liable to sprain his
elbow with the sudden jerk which is required for
that feat; or if a person takes to tumbling and
jumping, without proper training, he will probably
suffer for his temerity. Again, common experi-

ence tells us that a joint which has been sprained is, for a long time, liable to be sprained again. This is because the part remains tender as well as weak ; and the muscles do not brace it steadily and firmly, or come nimbly to its aid when it is in danger.

In consequence of the foot, in walking, being placed upon the ground with the toes slanting a little outwards, the *outer* and hinder edge of the heel first touches the ground. Hence this part of the heel of the shoe is usually worn down before the remainder. The ball of the little toe next comes to the ground, and the balls of the other toes follow in quick succession; and it is from the great toe—that is, from the inner side of the foot—that the last impulse is given to propel the body, forwards, over the other foot. In order to give full effect to this final impulse an especial muscle, the "Long Fibular" muscle (I in fig. 13), is provided. The tendon (i) of this muscle passes, behind the outer ankle, beneath the sole of the foot, to the great toe. It has the effect of pressing the ball of the great toe upon the ground, while it raises the outer ankle, and so contributes to throw the weight, across, in the direction of the other foot.

Thus the foot revolves upon the ground, from the heel to the balls of the toes, and from the outer edge of the former to the inner edge of the latter; and during the revolution, which has been compared, though the comparison fails in many points, to the revolving of the segment of a wheel, the ankle is raised' and advanced forwards.

On the complete and steady execution of this movement good walking chiefly depends, more particularly upon the full performance of the last stage of the process, viz. the rising fairly upon the balls of the toes and delivering the weight steadily over to the other foot. This is the most difficult part of the process, the whole weight of the body during its execution being borne upon the fore part of the foot, that is, upon the longer pillar of the plantar arch; forasmuch as the heel is being raised and the other foot is off the ground. For the good performance of this part of the process, all the features of the well-made foot are essential. There must be a high and firm plantar arch, a heel set at a proper angle, and a strong great toe running straight forwards. There must be also a fully developed calf to set the machinery well in motion.

If the plantar arch be low it cannot bear the strain attendant upon this movement; and the person, in consequence, shirks the full performance of it. He does that by turning the toes too much out; and, then, he contrives to roll over the inner side of the foot, instead of rising upon the balls of the toes ; and so he gets along with short, shuffling, feeble steps. How many persons, owing to one cause or other, hobble in this way! Some turn the toes very much in, and rise over the ball of the little toe, instead of over the great toe. This is done with comparative ease, because the ball of the little toe is nearer to the ankle ; but the step is, thereby, shortened, as well as rendered less firm and less graceful.

The revolving movement of the foot, or the bringing of its several parts into contact with the ground in succession, in a distinct manner, is peculiar to man. Many animals do not bear upon the heel at all ; they only tread upon the toes, and are, therefore, called DIGITIGRADE. Some, indeed, bear only upon the tips of the toes, as the Horse (fig. 5, p. 15). Others go upon the balls of the toes, as the Cat, the Hare, the Pig, and the Dog. Some animals bear upon the heels as well as the toes, and are called PLANTIGRADE, as the Bear, the Badger, and the Monkey ; but these all flop the sole upon the ground in its whole length at once. The foot in them is not sufficiently compact and strong to bear the weight of the body first upon one part then upon another; and they, consequently, walk in an ungainly manner, as compared with man.

Character evinced by manner of Walking.

Bear in mind that for the firm vigorous walk there is required, not only the well-formed limb, but also the manly and determined WILL, acting in a decided and authoritative manner over the several members of the body, so that these are accustomed readily, and steadily, to obey its commands ; just as the soldiers of a well-drilled regiment obey the directions of the superior officer in an orderly and efficient manner. And, as you may judge of the character of the officer by the discipline of his men, so may you form an estimate of a man by the movements of his limbs. You see a man walk along the street, and you instinctively form an opinion of him by the mode in which he carries himself and treads the ground. Be careful not to allow yourselves to be inflexibly biassed by these first impressions, as that amounts to prejudice. Nevertheless, experience tells us that they are not to be altogether despised. They originate in a perception of the working of the great laws by which body and mind are harmonized ; and, if fairly estimated, they rarely deceive us.

We have little difficulty in recognising three chief classes among pedestrians. *First,* there are those who pay too much attention to the movements, who walk with a pompous strut, or a mincing gait, or affect some style or other. We are naturally very little inclined in favour of such persons; indeed, we have usually to make an effort not to be decidedly prejudiced against them. *Secondly,* there are those who pay too little attention to their movements, who do not seem to be sufficiently alive to the responsibility attaching to the possessors of so noble a structure as the human frame, and who do not give themselves the trouble to exert the powers of the glorious mechanism with which they are charged. They slouch, or dawdle, along in a listless lazy manner. Instinct tells us, and tells us rightly, to beware how we trust such persons with the conduct of our affairs, or with any office of responsibility. We feel that the lack of energy manifested in the guidance of their limbs is, too probably, a feature of character, which unfits them for the active duties of life; and we know that such men are not usually successful in their calling. *Thirdly,* there are those who shew, by the firmness and precision of their step, and by the regularity in the succession of the movements by which the step is made, that they are conscious of the dignity of their species, of the responsibility attendant on that dignity, and of the respect due to themselves. Such men

we feel are likely to pursue their avocations ener-
getically and methodically, as well as with punc-
tuality.

Many points of character peep out in the way
men walk. Our poet tells us that in one we may
read

"rascal in the motions of his back
And scoundrel in his supple sliding knee."

Another has a halting, shuffling, undecided gait;
while a third walks in a bold, determined, straight-
forward, erect and independent manner. One
has a cautious, parsimonious step, as if sparing
of shoe-leather, or afraid to trust the ground; he
has, however, probably, trusted the funds with
considerable investments. Some walk with long,
pretentious, measured strides; others make short,
quick, insignificant steps. Some, again, are hur-
ried, fussy, noisy; while others glide along in a
quiet, shrinking, unpretending, it may be timid,
manner.

I need not dilate upon these diversities. Your
own observation will supply abundant illustra-
tions of the correspondence between character and
manner of walking.

The several movements in walking are under
the control of the WILL, and are directed by it,
to such an extent that the continuous agency of
the will is essential to the process. If the influ-

ence of the will be suspended, but for a moment, the action ceases, and the man falls to the ground. Nevertheless, the play of the individual muscles, and their co-ordination, or the manner in which their several movements are combined, are, in a great measure, independent of the will. They are, to a certain extent, automatic, and result from peculiar relations between the nervous and the muscular systems. The will may be compared to the driver of an engine, who, by turning on the steam, and maintaining the supply, sets the machine in motion, and regulates the rate of its speed; but the several wheels are so arranged that they go on irrespectively of his immediate superintendence. It would be impossible for the engine-man to attend to the working of each detail of his machine; and it would be too much for the will to have to direct all the movements of the limbs in walking. We should be wearied with such an effort of attention before we had walked across a room; for the exercise of the will is exhausting, and soon engenders fatigue. The more we think of any movement and take pains to direct it, the sooner we are tired and unable to continue it; and the more the attention is diverted, the less quickly do we experience a feeling of exhaustion; while those movements in the body which are not at all under the influence of the will—the movements of the heart for instance —go on unceasingly, through a long life, without any sense of weariness. What so prevents fatigue, when we are walking, as the diverting conversation of an agreeable companion?

But though the combination of the movements in walking is, to a certain extent, automatic, it is not complete without the proper control of the will. This is proved by the gait of those unfortunate beings in whom the mind, and with it commonly the will, is deficient from birth—I mean IDIOTS. Their movements are, usually, more or less, irregular and unharmonious, jerky, without proper steadiness and rhythm; the head is tossed about; the eye looks one way; the fingers are sprawled out in another direction; the foot is jerked out at a hazard, as it were, so that you don't know when it will reach the ground, perhaps it kicks against the other foot. A sad spectacle this. The visit to an Idiot Asylum fills one, it is true, with a sense of the value of an institution where these poor members of the human family are kept out of harm's way, and away from the gibes of the village boys, and are made clean, and tidy, and taught so far as they are capable of instruction; but I know no sadder sight than is presented by a string of the inmates of such an asylum, guided from room to room by the foremost of the number, who shews by his walk, somewhat more steady than that of the others, that he is gifted with rather more intelligence than they, and is so fitted to be their guide.

An equally melancholy, an even more distressing, spectacle is that of criminals pacing, like animals in their dens, up and down the court-yard of their prison; for in them we know, that there is no deficiency of will. It is strong enough to control and regulate the movements of their limbs; but there is a still more important deficiency, viz. a deficiency of that moral sense which should control the will.

Another sad, but physiologically interesting, sight is the rolling walk of the drunkard. Here, again, the will is not deficient; but it is, partly, and by its own agency, dethroned. Enough of the will is left to set the machine going, not enough to guide it and control it well. Though the movements follow one another, for the most part, in proper sequence, they are uncertain and ill-directed. The balancing power is partly lost. The feet are dragged hither and thither, and thrown about, by the swerving weight of the body; and they follow one another upon the ground at uncertain intervals, and in any but a straight line. You watch a man in this state staggering from side to side, and wonder how he keeps his legs at all. Soon the foot catches against some slight obstacle or against the other leg, or fails to come quickly enough into the required place, and the man rolls over. The supple manner in which his unstrung limbs give under the weight, perhaps, saves him, to some extent, from the shock; but you must not imagine that drunkards have any charm against injury. A large proportion of the accidents admitted into our Hospitals are the result of drunkenness.

Distinctive Features of the Human Foot.

I have already made a few comparisons between the human foot and that of certain of the lower animals. It will be interesting to add some others.

There are several animals, as the Monkey, the Bear, and some Reptiles, in which the foot resembles the human foot in many particulars. It has, for instance, the same number of toes as the human foot, and the same, or nearly the same, number of bones, and the latter disposed in much the same manner. Certain peculiarities, however, distinguish the human foot. These all have reference to the power which man, and man alone, possesses of standing firmly upright, and of walking steadily, upon the two feet.

The following are the most important of these distinctive features.

First. The several parts are fitted and bound together in a compact firm manner, so as to combine strength and elasticity in the highest degree. In this respect the human foot contrasts very remarkably with the sprawling foot of the Seal or Lizard (figs. 2 and 3, p. 11). The result is obtained, partly, by the great size of the tarsal bones, in proportion to the other components of the foot, and, partly, by the formation of the "Plantar Arch," which is higher and stronger in man than in any of the lower animals.

Secondly. The TOES are short and small in relation to the other parts of the foot. In many animals, the Monkey for instance (fig. 44, p. 89), the toes form the greater part of the foot; and, in some, the bones of the instep are reduced in number as well as in size : the reason being that, in such animals, the toes are required to perform a variety of offices—burrowing in the ground, scratching, holding on to the branches of trees, catching and tearing prey, &c.—for which their services are not needed by man.

It may here be noticed that one of the great points of dissimilarity between the foot and the hand consists in the difference which the length of the digits bears to the other components in the two members. They form nearly *half* the length of the hand, but not more than a *tenth* of that of the foot. Clearly, therefore, they constitute a far less important segment of the lower limb than they do of the upper, and are intended to perform much less important functions in it. In the hand the fingers and thumb may be said to constitute the essential part; whereas the toes do little more than help the foot to adapt itself to inequalities of the ground and so to obtain a firmer holding. In civilized countries, accordingly, where we walk, chiefly, upon even paths and paved streets, very little evil results from the loss of the services of the toes which is incurred by covering over the foot to protect it against the hardness of the roads.

We often hear the toes spoken of as ill-treated members, which are not allowed fair play because the art of man keeps them in a state of inertness and deprives them of their natural functions. Anatomy, too, gives some countenance to the idea, inasmuch as it shews that the muscles which minister to the toes are as numerous as those which are concerned in moving the fingers; and we occasionally see persons, who, having been born without hands, or having lost them, contrive to write and paint and do other unusual offices with their toes. Watch the movements in an infant's foot as yet unshod. They are considerably more free than in your own; especially you will observe that there is a power of separating the great toe from the others and approximating it to them which you have, probably, altogether lost. The small size, however, of the toes, and the comparative fixedness of the inner, or great toe, prove, that they were never intended for anything like the same variety of purposes as the fingers, and shew that, under the most favourable circumstances, the *pes* could never be *altera manus*, as some would persuade us that it is. Certainly it was never intended to be an organ of prehension. Hence, although in practice, boot-makers may excite our wrath and deserve our condemnation, I don't think that, in principle, they are so much to be complained of.

The *third* striking peculiarity of the human foot is the size of the inner or GREAT TOE and the firm manner in which its metatarsal bone is joined to the other bones, so as to render it a main pillar of support to the foot. These features of the great toe have reference to the share of the weight of the body which is borne by the inner side of the foot, more particularly during the last stage of the step, when the body is propelled forwards over the other foot. Hence it is sometimes called the "hallux," from a Greek word (ἀλλομαι) signifying to bound or spring. The *mobility* of the *thumb*, enabling it to be opposed so easily to each of the other fingers, is a characteristic of the human *hand;* and the *solidity* of the *great toe* is equally, or even more, characteristic of the human *foot.* The great toe should be continued, from the instep, straight along the inner edge of the *foot,* or inclined a little *in*wards; often, as before mentioned, its phalanges become inclined *out*wards so as to interfere with the other toes[1].

Though, in many animals the number of the toes is the same as in man, this is not the case in all; and we may trace a gradual and progressive diminution of the number, in the following order.

I have said (page 10) that the inner toe is incomplete in all animals, forasmuch as, in none, does it possess the same complement of bones as do the other toes. You will not be surprised to find, therefore, that it is the first to be missing. The ELEPHANT goes upon *five* toes; but if you look closely you will perceive that the inner toe (fig. 39, I.) has not attained even its usual incomplete number of bones. It is short of one; and the inner wedge-bone, which looks like a metacarpal

---

[1] In ancient times warriors were wont to cut off the *great toes* as well as the *thumbs* of their captives to disable them for further service (Judges i. 6, 7).

bone, is prolonged, downwards, to supply the place,

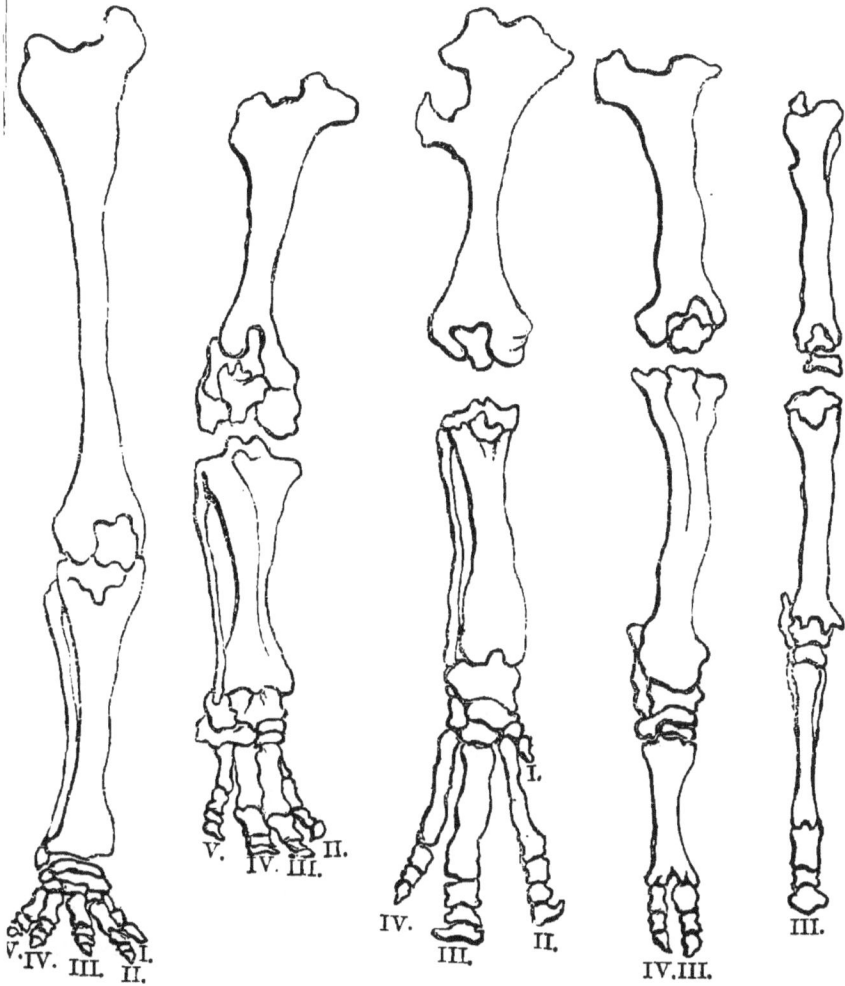

Figs. 39.   40.   41.   42.   43.
Elephant. Hippopotamus. Rhinoceros.   Ox.   Horse.

and to give sufficient length to the toe. The same
thing may be seen in some other animals, and it

is interesting as shewing the first indication of departure from what may be called the standard number of the phalanges. In the HIPPOPOTAMUS (fig. 40) we have an additional stage of imperfection in this same toe; for here there is only one small bone to remind us of the existence of the toe (it is the same in the Rhinoceros, I.); all the rest have failed to be developed; and the animal, consequently, goes upon *four* toes. Next the failure appears on the *outer* side of the foot, and affects the little toe. Thus, the RHINOCEROS (fig. 41) goes upon *three* toes—namely, Nos. II. III. and IV. —and there is scarcely a trace either of the first toe or of the fifth. In Ruminating animals, as the Ox (fig. 42), the second toe is wanting, as well as the first and the fifth; so that the foot rests upon *two* toes (Nos. III. and IV.); and in the HORSE (fig. 43), as we have already seen, only *one* toe—the middle one (No. III.)—is developed sufficiently to reach the ground.

Whatever pretensions to Humanity the MONKEY may make—and they are sufficiently striking to render some persons very uncomfortable on the score of relationship—he is certainly far removed from us in the construction of the foot (fig. 44); and the good people to whom I have alluded may derive consolation from the reflection that, in this respect at least, there is very little indication of

cousinship. Indeed we ought not to speak of his
foot at all; for the part which corresponds with the

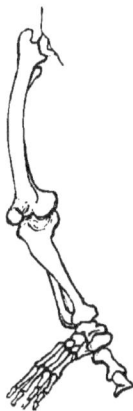

Fig. 44. Gorilla.

human foot does not even deserve that name. It
is so much more like a hand, that the term four-
handed, or *quadrumanous*, is by naturalists applied
to this class of animals. There is scarcely any
plantar arch; the animal bears, chiefly, upon the
outer edge of the foot; the digits are long and
strong; and the inner one, instead of being parallel
with the others, diverges from them so as to constitute a true *thumb* instead of a great toe. All these
points are very suitable for enabling the animal to
cling to branches of trees, and for other prehensile
purposes; but they unfit him for the upright posture, and render it impossible for him to walk
steadily upon his lower limbs.

In the great ape called the GORILLA, which is
found in the south-western part of Africa, and of

Fig. 45.   Gorilla.

which many specimens have now been sent to this
Country, the *hind-hand* is of great size and strength,
as may be seen in the accompanying drawing made
from a stuffed specimen in the British Museum.
The lower part of the leg is also very thick,
owing to the size of the muscles which move the
great toe and the other digits, and which enable
them to give a most powerful grasp.   So strong
and savage is the creature that all efforts to cap-
ture one alive, when full-grown, have, hitherto,

failed. He is said to give evidence of his strength of hand and of his amiable propensities in the following way. He swings by his fore-hands from the trees, and, letting himself down quietly by them, watches an opportunity of seizing by the neck, with his huge hind-hand, some unwary Negro who may be passing by, draws him up, and holds him with vice-like grasp, till his struggles have ceased, and then drops him a strangled corpse to the ground.

Most of the characters above mentioned as distinctive of the human foot—such as its compactness and strength, the height of the plantar arch, the shortness of the toes—are, like the size of the calf, most marked in the higher members of the human family, in those, that is to say, who are gifted with the highest intelligence. Thus the formation of the foot is found to have a correspondence with the formation of the head, and may, like it, be, to a certain extent, taken, as I have before remarked, to be an index of intellectual, as well as of physical, capacity. The relation between the intellectual power and the physical conformation of man, which is here exemplified, and which is maintained throughout the frame, is a subject of extreme interest, and is one which has not attracted the attention of anatomists and ethnologists so much as it deserves.

To what secondary causes this harmonious adaptation of body to mind may be due, we cannot clearly tell; but we can see in it a provision for giving physical ascendancy to superior intellect. And it is most gratifying to be able to derive, as we may do, from this as well as from the observation of the past and the present, the assurance that the cultivation of the mind, provided its moral tone be preserved and proper sanitary precautions be taken, is not likely to be attended with any deterioration of the body. On the contrary, we have good reason to believe that the present civilized nations of the earth, with their higher mental culture, are inferior to none of their predecessors in the qualities of the body; surely soldiers never maintained a hand-to-hand struggle better than the victors at Inkermann; and we know that the civilized nations are physically superior to most of the uncivilized. We have good ground, therefore, to hope that the extension of education and commerce will be productive, on the whole, of an improvement of the physical condition of the species.

Sir James Emerson Tennent says that the Veddahs, or aboriginal inhabitants of Ceylon, use the foot in drawing the bow. They sit down, place the toe against it, and draw the string with the hand; and some of the American Indians appear

to have used both feet in the same way. These
Veddahs furnish a good illustration of the low physical condition which is usually associated with absence of mental culture. They are described as in
a singularly degraded state. "They have scarcely
any language, no knowledge of God, nor of a future
state, no temples, no idols, no altars, prayers, or
charms; and, in short, no instinct of worship, except it be some addiction to ceremonies, analogous
to devil worship, to avert storms, lightning, and
sickness. All presented the same characteristics of
wretchedness and dejection—projecting mouths,
prominent teeth, flattened noses, stunted stature,
and other evidences of the physical depravity which
is the usual consequence of hunger and ignorance.
The children were unsightly objects, entirely na-
ked, with misshapen joints, huge heads, and pro-
tuberant stomachs. The women were the most re-
pulsive specimens of humanity I have ever seen
in any country."

The Proportions of the Limbs.

A few years ago I took the measurements of numerous skeletons which I found in the museums in France, Germany, and England, and made the following table to shew the proportions of the several parts.

The length of the foot and hand is in all somewhat greater than it should be, in consequence of the bones composing them being usually less closely articulated in the artificial skeleton than they are in nature.

From this it appears that the limbs of MAN differ from those of the APE, chiefly, in the proportionate length of the thigh and arm, and in the shortness of the foot and hand. And it will be seen that, in both these particulars, the NEGRO differs from the EUROPEAN and exhibits some approximation to the APE.

I found, also (the tables shewing this are given in my work on the Human Skeleton), that these characteristic proportions of the European are brought out only during growth; for that in the early periods of infancy the foot and hand are, relatively, very long, and the thigh is actually shorter than either the leg or the foot, and the arm is shorter than either the forearm or the

MEASUREMENTS OF SKELETONS (IN INCHES).

	Height	Middle point of.	Spine, length of.	Circumference of Skull.	Humerus.	Radius.	Hand.	Femur.	Tibia.	Foot.	Pelvis. Trans. diameter.	Pelvis. Ant.-post. diameter.
European (average of 25)	65	Symphysis pubis.	22.2	20.5	12.7	9.2	7.3	17.88	14.4	10.6	5.2	4.3
Negro (average of 25)	62	1 inch below Symphysis.	19.3	19.8	12.1	9.4	7.7	17	14.4	11.11	4.6	4.1
Bosjesman (average of 3)	54	Symphysis.	17	19.6	10.8	8.3	6	15	12.9	7.5	4.4	3.5
Idiot (in Berlin Museum)	57		19.5	13.5	12	8.8	7	16	12.5	8.5	5	3.8
Chimpanzee (average of 4)	50	3 inches above Symphysis.	17		12.2	11	9	12.4	10	10.5	4	5.5
Orang (average of 2)	44	3½ inches above Symphysis.	18		14	14	10	10.6	9.2	12	3.8	4.5
Gorilla (average of 3)	58	4 inches above Symphysis.	21		16.6	12.9	9	13.9	11.3	12	5.7	7.3

hand; and it is only, gradually, during the advance to manhood, that the proper proportions are attained. So that the transient or immature condition of the human frame shews certain resemblances to the permanent Negro type and to that of the quadrumanous animals; and these resemblances become obliterated during further growth.

The accounts of travellers indicate that some other nations present great varieties in the proportion which the length of the foot and hand bears to the height. Bushmen and Hottentots are very diminutive, commonly under 5 feet in height; and their hands and feet are remarkably small and delicate, in which respect they differ from Negroes. Mr Bartram observes with regard to the Cherokees or Muscogulges—a tribe of North American Indians—that the women are, perhaps, the smallest race of women yet known, almost all under 5 ft.; and their hands and feet are not larger than those of Europeans of 9 or 10 years of age. He tells us, also, what is very strange, that the men of this same tribe are of gigantic stature, " a full size larger than Europeans," many of them above, and a few under, 6 ft.; but he says nothing of the size of their hands and feet. The hands and feet of the Patagonians are said to be very small. This may be contrary to what we might expect; but it accords with what I found to be the case in the skeletons of some Giants which I measured; for in all of them the feet and the hands were disproportionately short. It would seem, therefore, that, whether the stature of the individual be diminutive or gigantic, the foot and the hand, in either case, are, usually, less than their proper relative length. A greater number of accurate data are, however, necessary to enable us to generalise correctly upon this and other points of a like nature, or to decide what truth there is in the common remark, that a long foot in a child indicates a tall man.

In former times the parts of the human body were used as measures; and it was not uncommon to illustrate the tables of measures by drawings of the human body, with descriptions of the foot, palm, &c. One of the tables of the 16th century, derived in great part from the Romans and the Greeks, is founded upon the notion, which is not very far from the truth, that in the well proportioned man, the breadth of the palm is a 24th part of the whole stature, and the length of the foot a 6th part, and the length of the cubit —from the elbow to the end of the fingers—a 4th. The measures, however, varied at different times and in different countries, even though the names were the same. The latter have, in several in-

H.   7

stances, remained, though the definite measure which they now indicate is different from what it was, and differs from that of the part of the body from which the name was taken. Thus, our present foot measure (twelve inches) is considerably more than the length of the human foot.

The Skin of the Sole.

The SKIN of the sole is soft and yet very tough and strong. It underlies a thick pad of fat, which separates it from the bones and the plantar ligament. The fat is interwoven with fibres passing, through it, from the tissue of the skin to the bones and ligaments. It is, in this way, rendered very firm, though it retains much of the soft quality of fat ; and it forms an admirable cushion for receiving the weight of the body and defending from injurious pressure both the skin and the other parts of the foot. The fibres just mentioned bind the skin to the superjacent bones and ligaments, and hold it firmly to them, so as to prevent its being displaced from them in the movements of the foot upon the ground.

The accompanying woodcut shows that these connecting fibres are most numerous where there is the greatest pressure, viz. beneath the heel and

the balls of the toes. It shows, too, that they
take the direction at each of those parts which

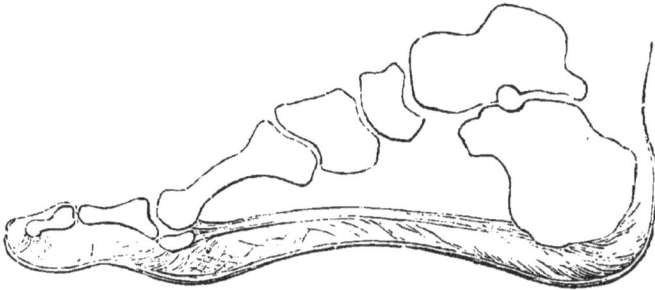

Fig. 46.

is most calculated to prevent displacement. Thus,
at the heel their direction is chiefly from the heel-
bone, backwards, to the skin. When we place the
heel upon the ground in walking, the weight of
the body has a tendency to drive the heel-bone
*for*wards from the skin; and the direction of the
fibres, from the heel-bone, *back*wards, just resists
this tendency and holds the skin and the bone
firmly together. On the contrary, when we with-
draw the foot from the ground the pressure is
in the opposite direction, and has a tendency to
drive the metatarsal bones *back*wards from the
skin. The course of the fibres is, consequently,
changed. They, many of them at least, run *for*-
wards from the bones and prevent the displacement that would be likely to occur. This direction is also very marked, and for the same reason,

at the end of the great toe. A bundle of fibres radiates from the projecting process, or tubercle, which is conspicuous upon the under surface of the bone near its end; and the greater number of them run *for*wards, through the pulp of the toe, to the skin, and maintain the connection of the skin with the bone when the latter is pressed *back*-wards in withdrawing the foot from the ground.

The skin of the sole has a peculiar sensitive-ness, which enables it to take quick cognisance of contact with the ground or of any injurious substances lying upon the ground. The sensitive-ness in the foot is rather increased by its being so much covered up. We are aroused to a con-sciousness of this sensitiveness when the soles are tickled, or when any one treads on our toes, espe-cially if there happen to be a corn there. We know also how sensitive the feet are to cold, and how liable we are to catch cold from wet feet. This sensitiveness renders washing the feet a re-freshing luxury, especially in hot climates or when we are fatigued. It is a luxury much indulged in by Eastern nations; "Mephibosheth had neither dressed his feet nor trimmed his beard from the day the king departed, until he came again in peace;" and among the Jews in our Saviour's time (Luke vii. 38), when guests were made very welcome, their sandals were unloosed, and their feet washed and carefully wiped, and, if the person were of high rank, anointed.

The integument of the foot varies in different animals, according to the nature of the ground upon which they tread and other circumstances. Thus the Elephant, the Hippopotamus, and the Rhinoceros, living in jungles and in marshy districts, have a more or less soft covering of skin. Oxen and Horses gallop about upon dry ground; and their feet are soled with thick hoofs of horn. The Dog has tough pads of skin with thick cuticle upon his feet; and the feet of the Feline tribe are muffled with fur so as to enable them to approach their prey with a noiseless tread. Man's foot is, by nature, like the rest of the surface of his body, comparatively unprotected; but as the foot, by its efficiency, emancipates the hand from the drudgery of carrying, so does the latter make some return for this relief by providing artificial coverings which enable the foot to tread upon various surfaces, and protect it against the inclemencies of the seasons.

On Shoes.

A few words on the subject of SHOES. No one
will dispute the correctness of the principle that
the shoe should be made to fit the foot; yet it is
not a little remarkable that this principle is so
often departed from in practice, and that the usual
plan is to make the foot adapt itself to the shoe.
That is, the shape of the shoe is determined according to the fancy of the maker or the dictates
of fashion, and the foot is expected to mould itself
accordingly. This is particularly the case with the
fore part of the shoe, into which the toes, or most
compliant parts of the foot, are squeezed. Thus,
the shape of the sole of a sound foot is about that
represented in fig. 47; the great toe is seen to be
free from the others, and the line of its axis, prolonged backwards, traverses the centre of the heel.
Compare this with the outline of the sole of a shoe
as usually made; and the violence that is done to
nature is at once perceived. The shoe is made
quite symmetrical, or is curved a little in the part
between the heel and the sole—in the "waist" as
it is called—when the shoes are to be worn on the
left and right foot respectively; and the toes, instead of being allowed to spread out a little, are

pressed together, and made to converge to a point in the line of the middle toe, as seen in fig. 48.

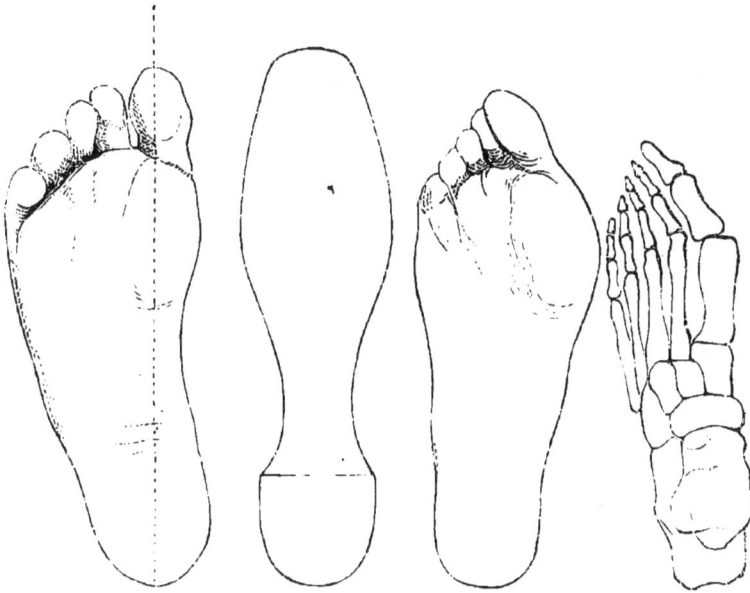

Figs. 47.   48.   49.   50.

The line of the great toe is thus quite altered, and the other toes are tightly wedged together (figs. 49 and 50); or, not being able to find room side by side, they overlap one another and form unsightly projections beneath the upper leather of the shoe. No wonder that "corns" and "bunions" and "in-growing toe-nails" are the frequent result of this treatment, and that so many persons are compelled to walk in a cautious, feeling manner, and to watch the ground narrowly, lest their cramped and tender toes come into contact with a stone or other projecting body.

How greatly to be lamented it is that the foot should be thus maltreated and distorted, and that walking should be made so painful, and that the shoe, which is intended to befriend and protect the foot, and which, if well fitted, would support it and preserve its shape, and make some amends to it for the rough hard roads upon which it is compelled to tread, should be thus perverted into a means of galling it and impairing its functions.

This subject has been treated of in a simple and concise manner by Dr Meyer, Professor of Anatomy at Zurich, in a small pamphlet, which has been translated into English by Mr Craig, and entitled, "*Why the Shoe pinches*[1]." I hope it may be read by boot-makers, and may lead to some improvement in their art. Dr Meyer very properly remarks that one of the main points to be attended to is, to allow the great toe to have its normal position; and this can be done by making the inner edge of the sole incline *in*wards, from the balls of the toes, instead of *out*wards. The accompanying

---

[1] *Why the Shoe pinches*, a contribution to Applied Anatomy by Hermann Meyer, M.D. Professor of Anatomy in the University of Zurich, translated from the German by John Stirling Craig, L.R.C.P.E., L.R.C.S.E., price sixpence.

The preceding four figures and the two following are taken from this pamphlet with Mr Craig's permission.

drawing (fig. 51) gives the outline of a shoe designed under his superintendence, and shows the

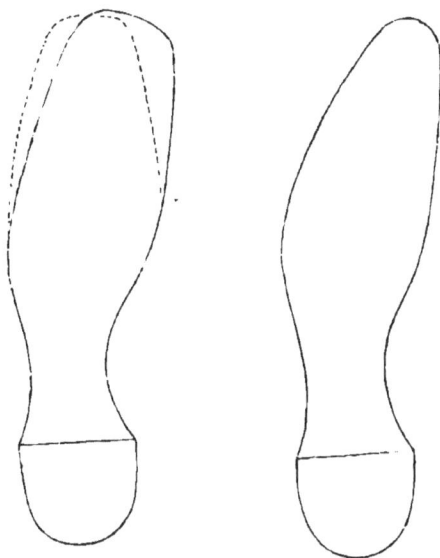

Figs. 51.   52.

difference between it and the usual shape, the latter being indicated by the dotted outline. In fig.
52 the shoe is pointed, the pointing being effected
from the outer side. I have often laboured, but
laboured in vain, to impress the same point, and
hope the more systematic attempt of Professor
Meyer may lead to better results.

With regard to the *heel-piece*, I have already
said that it should not be high because it makes
the step less steady and secure, and at the same
time shortens it, and impairs the action of the

calf-muscle. A high heel-piece, moreover, renders the position of the foot upon the ground oblique, placing the fore part at a lower level than the heel; thus the weight is thrown too much in the direction of the toes, and they are driven forwards and cramped against the upper leather of the shoe. The high-heel of a boot, therefore, tends to aggravate the evils which are caused by the insufficient and ill-adjusted space which is allowed to the toes.

This account of the foot has necessarily been very superficial and imperfect. There are many points in its anatomy to which I have not even alluded; but, if I have succeeded in giving you some idea of the general plan of its construction, and in stimulating you to further enquiry respecting the mechanism of the Human Frame, my purpose will have been served. Still more will it have been so, if you carry away with you some sense both of the Pride and of the Humility which the review of such a structure is calculated to excite—of pride, not selfish pride, but pride resulting from a consciousness of the nobility of your physical nature, a pride which will make you spurn what is bad and degrading, and will help you to aspire to what is elevated and good. The impressions resulting from a comparison of this one fragment of Nature's work with our own most laboured achievements must quell any other pride; and the very admiration with which we contemplate the structure of our body impels us to walk humbly with our God, whose gift that body is.

THE HUMAN HAND.

THE great characteristic of the Hand, as distinguished from the Foot, is the mobility of the first digit, or thumb. Accordingly when this digit stands out apart from the others, and can be moved independently of them, so as to be more or less completely opposed to them, in the upper or Mammalian Class of animals, at least, we call the member a Hand. When this digit is absent, or is fixed in the same manner as the others, which is the case in each of the four limbs of Quadrupeds, we call the member a Foot. In Monkeys, or in most of them, the thumb is present and is separate and moveable in each of the four limbs; and these animals are, therefore, called "quadrumanous" or "four-handed." Man, having the moveable thumb upon each of the two upper limbs only, is "bimanous" or "two-handed." By this peculiarity, perhaps more definitely than by any other, he is distinguished in structure from all the rest of the animal series; and naturalists have, accordingly, given the epithet "Bimanous" to the class in which he is placed, and in which he stands alone.

The hand is the executive or essential part of the upper limb. Without it the limb would be almost useless. The whole limb is, therefore, so made as to give play and strength to the hand; and, in ever so brief a description of the hand, it is necessary, even more than in the case of the foot, to give some idea of the manner in which the other parts of the limb are constructed, and to dwell a little upon such points as have relation to its movements.

The general plan of construction of the upper limb will readily be understood by means of the

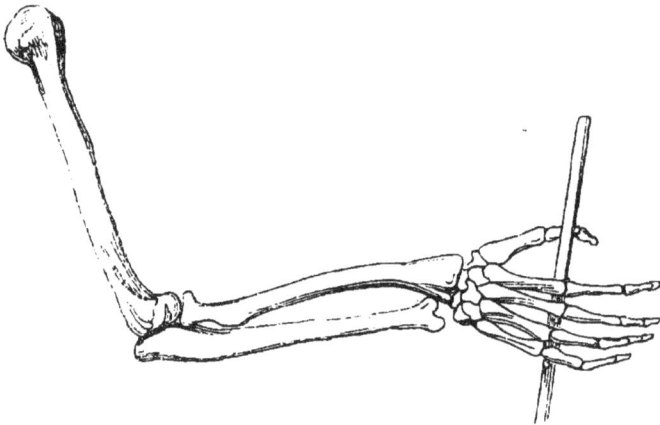

Fig. 53.

drawings (figs. 53 and 58, p. 122). It resembles

very much that of the lower limb (see fig. 4, page 15). The one bone of the upper arm—the *hume-*

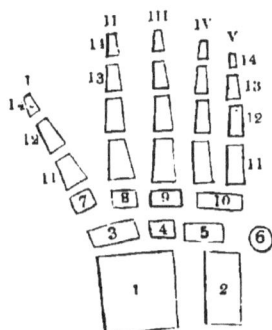

Fig. 54.

rus—resembles the one bone of the thigh, and is jointed, above, with the shoulder-blade, which, with the collar-bone, corresponds with the pelvis. Below, it is connected with the two bones of the forearm—the *radius* and *ulna;* and these correspond with the two bones of the leg. In the wrist there are eight bones, called *carpal* bones, arranged in two rows. These are connected with five *metacarpal* bones; and these, like the metatarsals of the foot, are jointed with the *phalanges.* Of the latter there are three in each finger; but in the thumb, as in the great toe (page 10), there are only two.

The diagram shows how the bones of the hand are arranged in three divisions. Thus, the upper row of carpal bones (3, 4, 5) consists, practically, of three bones; the fourth (6), which is much

smaller than the others, being rather an append-
age to one of them than a distinct constituent of
the wrist. (According to this view, the number of
the wrist-bones corresponds exactly with that of
the tarsal bones of the foot, viz. 7). The *outer* of
these three carpal bones (3) bears the thumb[1] and
the fore-finger (I. and II.), and constitutes, with
them, the outer division of the hand; the inner
one (5) bears the ring-finger and the little finger
(IV. and V.), and constitutes the *inner* division of
the hand; and the middle one (4) bears the middle finger (III.), and is the *middle* division of the
hand. The diagram shows, too, that the two outer
bones (3 and 4), with the two outer divisions of
the hand, are connected with the radius; whereas
the inner bone (5) only, with the inner division of
the hand, is connected with the ulna. Strictly
speaking, even this bone is not directly connected
with the ulna, but is separated from it, as will be
shown presently, by a thick ligament.

You frequently hear ignorant persons (and the
greater number of persons are lamentably ignorant
of the structure of their own body) speaking of the

---

[1] In deference to custom we call the palm the *front* of the
hand; and, therefore, we speak of the thumb as the *outer* and the
little finger as the *inner* digit: though it would better accord
with the ordinary position of the part, with its correspondence
with the foot and with comparative anatomy, to reverse these
terms.

small bones of the shoulder, or the *small bones* of the elbow. You may think this a matter of no importance, and that it does not concern you and people generally to have any knowledge of human anatomy. But I will tell you what is very often happening, and will leave you to judge whether such complete ignorance on this subject is not attended with some practical disadvantage. A man meets with an injury, falls and hurts his shoulder. The immediate effects of the injury subside; but he does not quickly recover the use of the part; he still cannot raise his elbow, or put his hand upon his head, or put it behind him. Soon he begins to think that something more is wrong than has been suspected; and the notion creeps over his mind, and gradually takes possession of it, that some small bone is displaced. Not content with the assurances of his medical man, he resorts to a quack, called a "bone-setter." The latter, taking advantage of the popular fallacy, gratifies the patient with the information that his fears are correct, affirms that "a small bone is out," and proceeds forthwith to employ the requisite forcible measures for putting the said "small bone" in. I need not say with what result. Every year, in this civilized country, many persons are maimed for life by these attempts to put imaginary small bones in. I beg you, therefore, particularly to ob-

H. 8

serve that *there is no small bone* either at the shoulder or at the elbow. The only small bones are at the wrist; and these are so well fitted to one another, and so firmly bound together, that nothing short of a crushing force suffices to displace them. This remark respecting the small bones of the wrist is true of nearly all the small bones in other parts of the body. So that, in fact, small bones are very rarely dislocated; and when you hear it asserted that a small bone is out, you may pretty confidently conclude that the speaker does not know what he is talking about.

I have said that the upper limbs resemble the lower in their general construction. There are, however, some important differences; and one of the chief of these is the greater variety and freedom of the movements in the upper limbs. *Strength,* for the purpose of carrying the body, is the object in the lower limbs. *Mobility* is the requisite in the upper limbs. Of this one example has already been given in the instance of the thumb as compared with the great toe.

Movements at the Shoulder.

An equally striking example is afforded by the shoulder. In the first place, the " Shoulder-blade" itself can be moved in several directions—upwards, downwards, backwards and forwards;—whereas the "Pelvis," i. e. the part which bears to the lower limb the same relation that the shoulder-blade does to the upper-limb, is immoveably fixed.

Secondly, the "Shoulder-joint" is so made as to permit a great variety and extensive range of movements to take place. We can move the arm forwards or backwards, as in throwing a ball, or, in sword exercise; we can raise it so that the limb points straight upwards; and we can swing it round in any direction. It is owing to the free movement in this joint that we are able to apply the hand to every part of the body, so as to remove sources of irritation. It is interesting to observe how other animals get on without hands, though they are much exposed to what we should consider great annoyance, as from flies, &c. The Cow, for instance, lashes its hide with its tail. The Cat licks itself with its tongue. The Sparrow dusts itself by the road-side. The Pig and the Donkey roll in the mud. And many of them, as the Horse and the Ox, have a thin muscle, called "panniculus carnosus," spread out under the skin, which effects those sudden twitchings of the skin whereby they are enabled to jerk off anything that troubles them. In Man the hand answers better than all these methods combined ; and it is necessary that it should do so, because his skin is more sensitive and less protected by natural covering than that of any other animal.

For this freedom of movement of the arms, so important to the usefulness of the hand, we are much indebted to the "Collar-bones." These bones, so called because they are placed at the

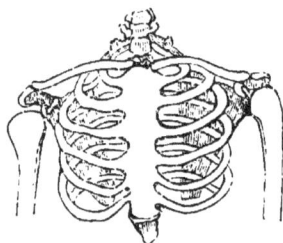

Fig. 55.

Chest and shoulders of man.

lower part of the *collum* or neck, extend, horizontally, from the upper edge of the breast-bone, to the processes of the blade-bones which overhang the shoulder-joint. Thus they hold the shoulders apart, and give width to the upper part of the chest. They also steady the shoulder-blades, and afford a *point d'appui* to the muscles which effect the lateral movements of the arms,—for instance, to the muscles which tend to draw the arms together, as when we hold anything, between the hands, in front of us; and to those which separate the arms from one another, as when we stretch them out at right angles with the body.

Many animals—the ELEPHANT, the RHINO-
CEROS, the HORSE and the OX—have no collar-
bones ; and they are only able to swing their fore
limbs to and fro. They cannot execute any lateral movements. They cannot throw the limbs
out sideways, nor press their fore feet together,
so as to hold anything between them. If the
horse wants to seize or hold any substance he
must do it with his mouth. The Elephant has
a special provision for the purpose of prehension
in his trunk, which enables him to provide himself with food by pulling down the branches of
trees. The LION and the TIGER can press their
fore paws together sufficiently to enable them to
hold their prey, and fix it upon the ground, while
they put the head down to it and pull at it and
tear it with their teeth ; and they are furnished
with rudimentary, or half, collar-bones suspended
in the flesh of the upper part of the chest ; while
the little SQUIRREL, which sits upon its hind legs,
and holds up the nuts between its fore paws to be
nibbled, has complete collar-bones. So has the
flying BAT, the climbing SLOTH and the digging MOLE. In BIRDS the collar-bones (fig. 56, AA)
are very large ; and, for the purpose of giving
them greater strength, they are united together
in the middle line just above the breast-bone,
forming what is commonly called the "merry-

thought;" and, as this is not sufficiently strong
to resist the force of the powerful muscles which
flap the wings and sustain the animal in the air,
there are, in addition, stout "side-bones," called by
anatomists " coracoid bones." These (B) run, from
the breast-bone (D), in the same direction as the

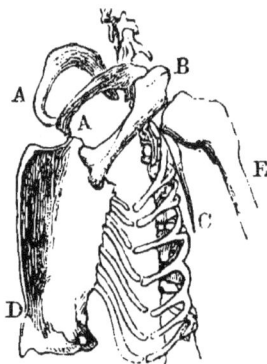

Fig. 56.

Chest and shoulders of bird.

collar-bones, one, on either side, to the shoulder-
blades (C); and they afford even more efficient
support to the shoulders than do the collar-bones.
The coracoid bones are peculiar to oviparous
animals, or nearly so. In some reptiles, as the
CROCODILE, they quite supersede the collar-bones.

These few examples are enough to show that
freedom of movement of the arms, especially of
lateral movement, is closely associated with, and,
indeed, is dependent upon the shoulder-blades

being supported and steadied by bones, which extend from the breast-bone to the shoulder-blades, and fasten the one to the other.

But, even the powers and advantages conferred by nature have often some drawbacks; and this free play of the arm at the shoulder in man, of which we are speaking, and the provision for it afforded by the collar-bone, are no exceptions to the remark. It is necessary for so great a range of movement that the socket in the shoulder-blade should be shallow, and that the ligaments which connect the arm-bone with the blade-bone should be loose. Hence the shoulder-joint is weak as regards its ability to resist injury. The collar-bone also causes the shoulder to project so much that it is greatly exposed to injury and often bears the brunt of a fall. A man is thrown from a horse or is knocked down upon the ground, and, if anything prevents the hand being stretched out, the chances are that he falls upon the shoulder. True, the head is saved thereby; but the shoulder suffers. Hence the shoulder-joint is more often dislocated than any other; and no bone is more frequently broken than the collar-bone. Even in little children, in whom, notwithstanding their many tumbles, the other bones usually contrive to escape, the collar-bones are often broken; and in grown-up persons the shoulder is sometimes

dislocated by the mere action of the muscles, as
in swimming, or throwing, or lifting a weight
above the head.

That you may understand the movements of
the shoulder a little more fully, I will ask you to
contrast the drawing (fig. 58), which shows the
position of the blade-bone upon the chest in Man,
with the drawing (fig. 57) of the corresponding

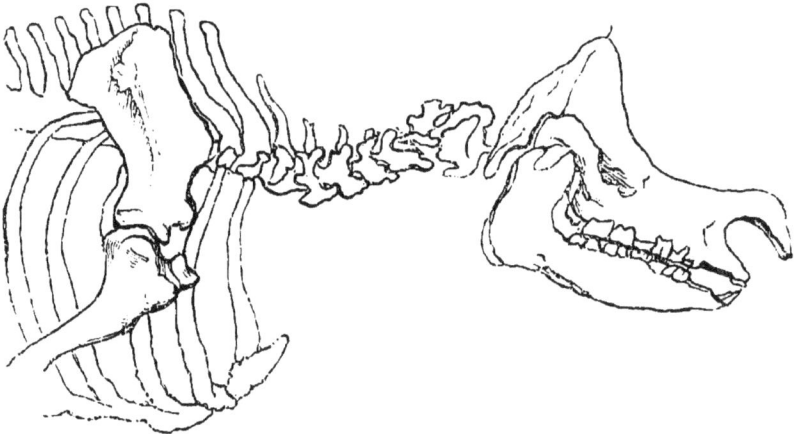

Fig. 57.   Rhinoceros.

parts of the Rhinoceros; and you will at once
recognise several important differences, besides
the presence of the collar-bone in the one and
its absence in the other.

In the RHINOCEROS the chest is deep, from
the back-bone to the breast-bone, and is flattened
at the sides; and the depth of this part of the
trunk is increased, slightly, by the breast-bone
projecting, keel-like, underneath, and, much more,

by the spines of the back-bone running up into a high ridge, above. The blade-bone and the arm-bone are applied against the flat side of the chest, and lie, lengthways, between the spine and the breast-bone, nearly parallel with the broad flat ribs. The blade-bone has no process overhanging the shoulder-joint, and, as before said, there is no collar-bone. The short thick arm-bone descends nearly in a line with the blade-bone, and has huge processes at its upper end for the attachment of muscles. The parts are designed to bear the great weight of the animal, and to carry its ponderous head and horn ; but the only movement of which they admit is a sliding of the blade-bone and arm-bone, backwards and forwards, upon the side of the chest.

In animals of similar construction to the Rhinoceros, but of lighter frame, and of greater fleetness, the blade-bone is placed more obliquely, which gives freer and easier movement both to it and to the arm-bone. This, for instance, is the case with the well-bred horse, and if we want a quick-going horse, one that can lift his fore feet well, we should observe whether the shoulder-blade is oblique, and whether the spines of the back rise well above it. Such a horse is said to have "a good shoulder" and to be "well up." He will carry a saddle well, and is not likely to trip.

In MAN the chest has proportionately less depth and length, and greater breadth, than in any other animal; the breast-bone is quite flat; and the spines of the back are sloped downwards, so that they do not project beyond the level of the ribs and the blade-bones. Hence he can lie

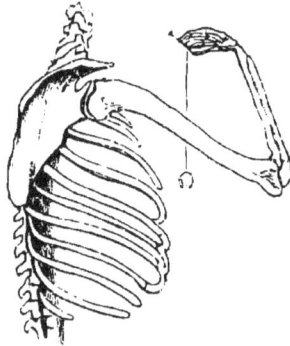

Fig. 58.

easily either upon the stomach or the back—a privilege which is shared with him by very few of the lower animals. Scarcely any of them can lie upon the back, or even upon the stomach without the help of the fore limbs. The donkey enjoys rolling over and over upon a dusty road, but he cannot poise himself for a minute upon his back.

The sides of Man's chest, moreover, are not *flat*, as in the Rhinoceros and Horse, but *rounded*, so that the blade-bones can revolve upon them to and fro, as well as slide upwards and downwards; and the long arms—comparatively long, that is, from the shoulder to the elbow—hang quite free of the chest and form sharp angles with the blade-bones.

The blade-bones are accommodated to the shape of the chest; for, instead of being elongated in a direction parallel with the ribs, they are prolonged downwards, along the sides of the chest, at right angles with the ribs. This prolongation of the lower part of the blade-bone is very important, inasmuch as it enables the muscles to hold the bone steady upon the wall of the chest, and so gives greater power to those muscles which pass from the blade-bone to the arm and act upon the shoulder-joint. Were it not for this provision the contraction of the muscles intended to raise the arm would quite fail to produce the desired effect, and instead of it would simply cause the shoulder-blade to revolve upon a transverse axis. That is to say, when we endeavoured to raise the arm our effort would merely have the effect of raising the hinder part of the shoulder-blade.

In each of these particulars—in the form of the chest, and in the shape and direction of the shoulder-blade—the Monkey is intermediate between Man and the inferior animals. The Monkey's chest is broad and round, in proportion

to its length, if we compare it with other animals;
but this is less marked than in the human chest.
And you perceive that the Monkey's back-bones
project, as they do in other animals, beyond the
level of the ribs. The blade-bones are also like
those of Man in being prolonged downwards, and

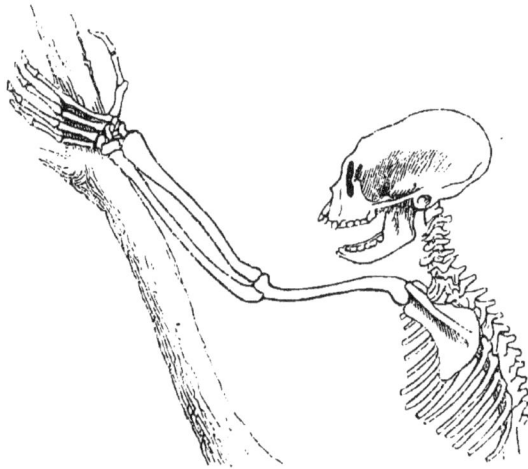

Fig. 59.   Monkey.

in being carried, to a certain extent, across the
ribs; but their lower angles do not run so far
in this direction as they do in the human skeleton.

The movement of raising the arm, as in carrying the hand *outwards*, or pointing upwards, or
putting the hand upon the head, is rather a
difficult one, and requires the combined action of
many muscles. It is, therefore, to be avoided by

persons to whom muscular straining is likely to be injurious; and the power of effecting this movement is easily impaired by accident or disease. A long time often elapses even after a slight bruise of the shoulder, before the person recovers the power of putting the hand upon the head.

The exercise of raising the arms above the head is a good one for those in health, and is much, and wisely, recommended by the directors of gymnastics. It brings many muscles into play, not only those of the shoulder, but the muscles all round about the chest, viz. those which pass from the spine and ribs, as well as from the breast-bone, head, and pelvis, to the shoulder-blade and arm; and, thus, it tends to strengthen the spine and the chest, as well as the shoulders and arms. There is, perhaps, no exercise so good as this; and it is much to be regretted that the dress of young ladies, with its paraphernalia of stays and shoulder-straps, interferes so greatly with it. The frequency among them of "pigeon-breast" and "crooked spine" must, partly, be attributed to the confinement of the arms, caused by the mode of dress and the customs of life. One of the few opportunities afforded to the arms of availing themselves of this exercise is in the dressing-room during the process of brushing the hair. I would by all means, therefore, recommend young ladies to give sufficient time and attention to this part of the toilette, and not to delegate it to the lady's maid. If, in addition, I suggest that it be commonly done with open window, I feel sure that I shall have a deservedly great authority among them—Miss Nightingale—on my side.

The movement at the ELBOW is, merely, that of bending and straightening, in a hinge-like man-

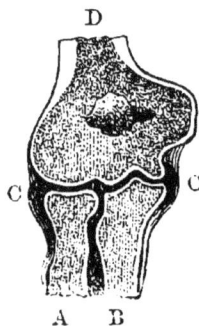

Fig. 60.   Elbow-joint.

ner; yet there is a slight obliquity in the direction in which it takes place, an obliquity resembling that in the movement at the knee (page 39).

Pronation and Supination of the Hand.

In the FOREARM and HAND there is a movement with which we have nothing exactly corresponding in the leg. It is called "Pronation and Supination." In *pronation* we turn the palm *down*wards, as in picking up any substance from a table; in *supination* we turn the palm *up*wards, as a boy does when he holds out his hand for a caning, or for the more agreeable purpose of having a shilling put into it.

PRONATION and SUPINATION take place in the following manner. Each of the two bones of the forearm extends from the elbow to the wrist (fig. 53); but one of them—the "ulna"—is chiefly connected with the elbow; and the other—the "radius"—is chiefly connected with the wrist, and, by means of the wrist, with the hand. The two bones are separate from one another, except at their ends. There they touch, and are jointed together in such a manner that the large lower end of the radius can play round, or partly round, the small, button-like, lower end of the ulna; and, in so doing, it carries the hand with it. In this movement the upper end of the radius (A, fig. 60) does not leave its place, but simply revolves, upon its own axis, on the surface of the arm-bone; and its edge turns in a notch cut for it in the upper end of the ulna (B), which remains still.

In the drawings (figs. 61 and 62) the relation of the parts in the supine and in the prone state is shown by the aid of a plumb-line falling from the part of the arm-bone upon which the upper end of the radius revolves. The line traverses the upper end of the radius, then passes along the in-

terval between the two bones, then traverses the
lower end of the ulna, and, finally, takes the

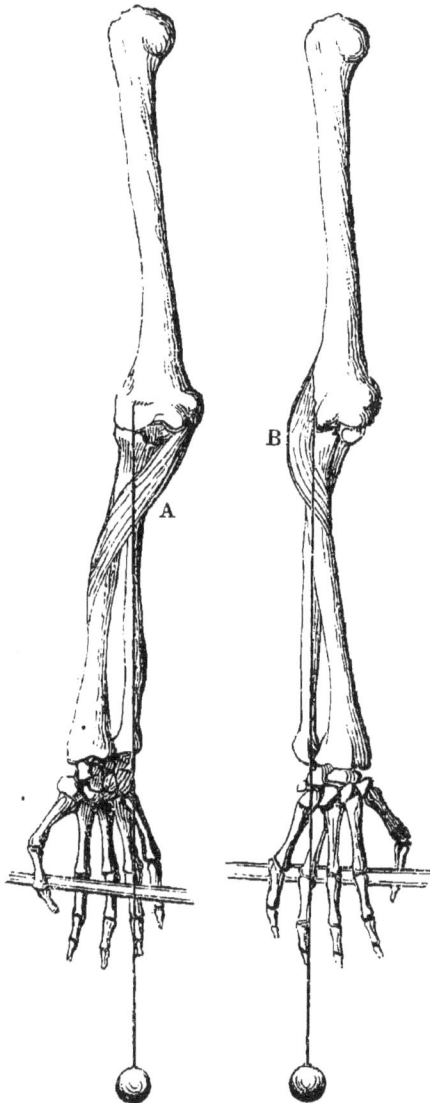

Fig. 61.

Hand supine.

Fig. 62.

Hand prone.

course of the ring finger. And, provided the limb be held vertically, the line traverses the same parts whatever be the position of the forearm and hand. It does so in complete supination, as shown in fig. 61; it does so in complete pronation, as shown in fig. 62; and it does so in every intermediate position. We may call it, therefore, the axis upon which the radius and the hand turn in pronation and supination; and, according to this representation, the ring finger remains stationary during the movement, while the other fingers and the thumb perform their partial revolutions around it.

I have said there is no movement in the lower limb exactly like the pronation and supination of the forearm and hand. We have, it is true, a power of moving the leg upon the thigh in a somewhat similar manner; but this can only be done when the knee is bent. For instance, when sitting in a chair with the foot upon a fender, or with the toes upon the ground, we can make the foot revolve so as to turn the heel in or out. A little careful observation, however, will prove that this movement takes place, altogether, at the knee, and that *both* bones of the leg participate equally in it, the *whole* leg revolving with the foot. Whereas, in the case of the forearm, the movement takes place, partly, at the wrist, and, partly, at the elbow; and *one* bone (the ulna) remains *still* while

H.   9

the lower end of the other bone (the radius) revolves around it. Moreover, the pronation and
supination of the hand and forearm are much more
free than these movements of the foot and leg;
and they take place with equal facility and freedom in any position of the limb. We can turn the
palm up or down as easily when the elbow is
straight as when it is bent.

The movement of which I am speaking is so
important to the usefulness of the hand, that I
will call your attention to three of the muscles by
which it is effected.

And, let me remark, by the way, that all the
movements in the solid parts of the body—probably all without exception, even the slight wrin-
klings of the skin that take place when it is exposed to cold—are the result of muscular action.
Muscles are bundles of fibres which have usually a
red colour and constitute what is commonly called
the "flesh" or "lean meat" of animals. They are
endued with the power of contracting or short-
ening themselves; and it is this property which
gives rise to the various movements of animal
bodies. At their ends muscles often dwindle into
"tendons" or "sinews" which, though occupying
much less space, and having no contractile power,
are very strong, and serve to connect the muscles
with the bones.

One of the three muscles just mentioned (A, fig. 61) passes from a projecting process on the inner side of the arm-bone, at its lower end, to the outer edge of the middle of the radius. Its contraction causes the radius to roll over, or in front of, the ulna. It thus pronates the hand, and is called a "*Pronator*" muscle. Another muscle (B, fig. 62) passes, from a projecting process on the outer side of the arm-bone, to the inner edge of the radius near its upper part. It runs, therefore, in an opposite direction to the former muscle and produces an opposite effect, rolling the radius and the hand back into the position of supination. Hence it is called a "*Supinator*" muscle.

The third is a very powerful muscle. It is called the "*Biceps*" muscle (fig. 63), because it has *two* points of attachment to the shoulder-blade. It descends along the front of the arm, and, bulging there, forms a conspicuous feature, to which athletic persons are proud to point in evidence of their muscular development. Its tendon crosses over the front of the elbow, and is inserted into the hinder edge of a stout tubercle which is seen on the inner side of the radius near its upper end. The chief effect of this muscle is to bend the elbow; but it also rotates the radius so as to supinate the hand; and it gives great power to that movement. When we turn a screw, or drive a gimlet, or draw a cork, we always employ the *su-pinating* movement of the hand for the purpose;

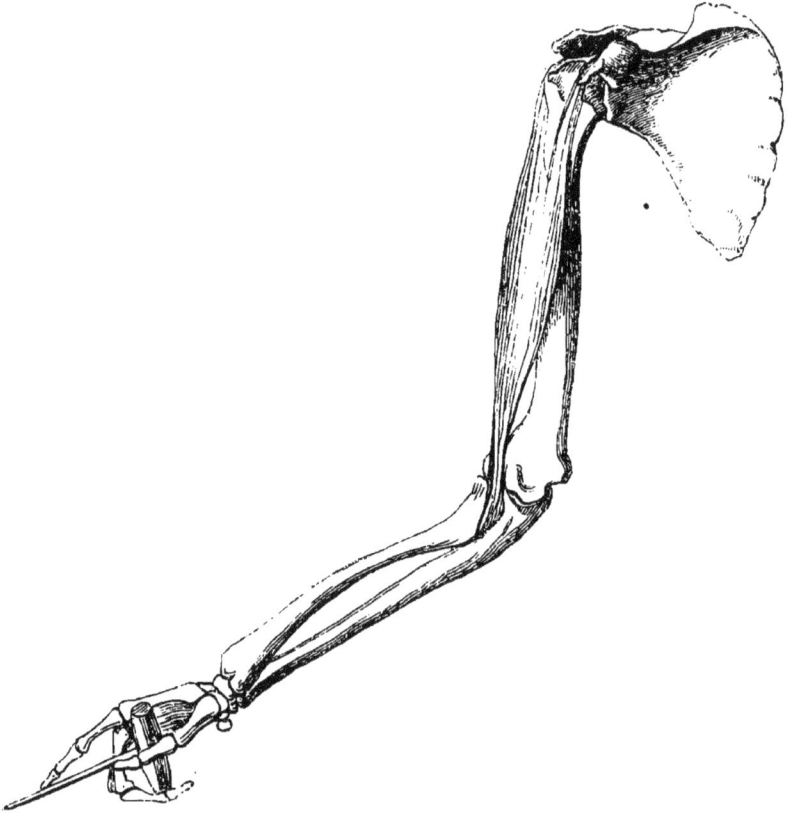

Fig. 63.   The Biceps Muscle.

and all screws, gimlets, and implements of the like
kind, are made to turn in a manner suited to that
movement of the right hand, because mechanicians
have observed that we have more power to supinate the hand than to pronate it, though they are,
probably, not aware that the preponderating influ-

ence of the *biceps* muscle is the cause of the difference.

The movement of which I am speaking is performed to its full extent only by Man. Monkeys cannot completely supinate the hand; and in most of the lower animals the part corresponding with the hand remains nearly, or quite, fixed in a state of pronation. Even in Man, complete supination is rather a constrained and awkward position. It is not a position which is habitual or natural to us. When we see any one sitting or walking with the palms turned forward it strikes us as strange, and the idea is suggested to us that the individual must be strange too, that, possibly, his head may be a little turned as well as his hands. In a state of ease the hand is naturally more or less prone; so that when it is desired to place the forearm or hand at rest, as in case of disease or injury, the prone position is usually selected. If the forearm be broken, for instance, the surgeon sets the fracture and fixes the limb with the hand prone or semiprone. This is, also, the position of greatest strength, as well as of most ease. Hence, in striking a blow, or carrying a weight, or making any strong muscular effort, the palm is always kept more or less inturned.

The Wrist.

This drawing (fig. 64) represents what is seen
when a section has been made, from side to side,

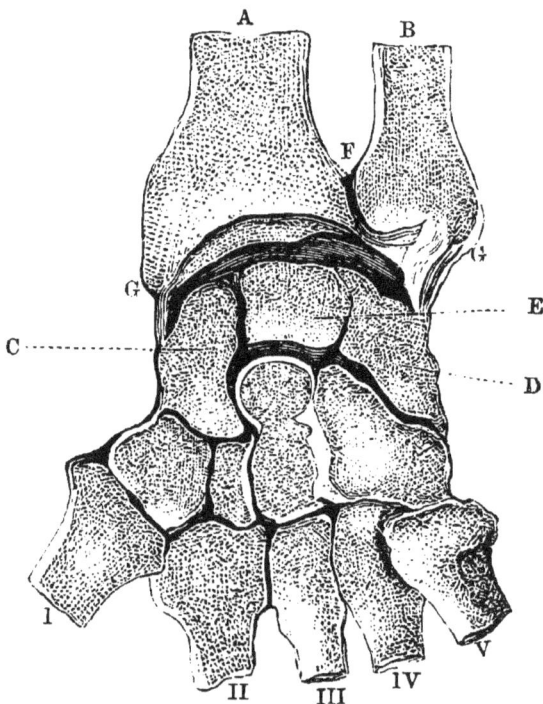

Fig. 64. Wrist-joints.

through the wrist and lower part of the forearm.
It gives an idea of the mode in which the several
bones of the wrist are adapted to one another and
held together by ligaments. The upper three
wrist-bones (C, E, D) are joined together, so as to
present a convex surface, which is received into a
wide cup, or socket, formed by the end of the

radius (A) and a ligament passing from the radius to the ulna (B) ; and, in pronation and supination, the end of the radius, together with this ligament and the wrist-bones, revolves upon the end of the ulna.

All the bones here represented are so well fitted to one another, and so strongly bound together, that, as I have before said, displacement very rarely occurs among them. We sometimes *hear* of a dislocation of the wrist, but very seldom *see* one. The wrist is often bruised, or its ligaments strained, by falls upon the hand; or, what very often happens, especially after the middle period of life, the bones of the forearm are broken a little above the wrist.

You might think that, in such an accident, the radius only would suffer, inasmuch as it is especially connected with the wrist-bones, and so receives the force directly from the hand. But, if you observe the line of contact of the radius and ulna (running from F), you will see that it is oblique, and that its direction is such as to cause the ulna to support the radius, and to receive some of the force from it; and this disposition, which makes the ulna share the duties of the radius, makes it, also, share the dangers; hence, it is very frequently involved with the radius in fracture of the forearm.

By the joints of the wrist we are enabled to move the hand backwards and forwards, and also slightly sideways.

The Movements of the Hand.

I come now to speak more particularly of the movements that take place in the Hand. I have already said that the mobility of the thumb is the chief characteristic of the hand as distinguished from the foot. Another important distinction between the hand and the foot is the greater length and mobility of the fingers as compared with the toes. The toes are short; and our power of moving them is, under any circumstances, slight. They constitute a small, and, comparatively, unimportant, part of the foot. The fingers, on the contrary, are long; they form a half, and, including the thumb, the more important half, of the hand. Without them the rest of the hand, indeed the rest of the limb, would be comparatively useless. Their movements are varied and free, and take place with singular facility and rapidity. We can bend them quite down upon the palm, and can extend them beyond the straight line; we can separate them from one another to a considerable extent; and we can bring them together with some force, as a waiter does when he carries a number of wine-glasses between his fingers; and persons who have lost the thumb contrive to hold a pen, a knife or fork, or other things, between the fingers.

Let me endeavour to give you an idea of some of the muscles which are concerned in executing these movements.

The wrist and hand are bent forwards upon

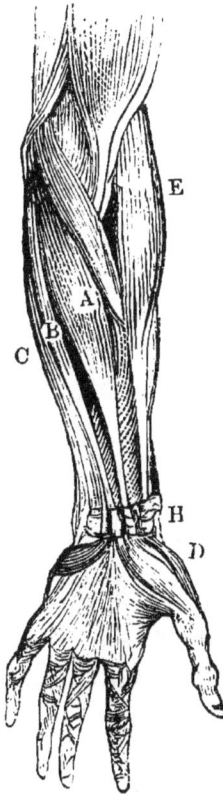

Fig. 65.

Muscles of forearm and hand.

the forearm by means of three muscles (A, B, C, fig. 65). These all pass downwards from the inner side of the lower end of the armbone. The outer and inner ones (A and C) are connected, by tendons, with the wrist-bones; and the tendon of the middle one (B) runs over the wrist and becomes spread out in the palm like a fan, so as to support the skin of the palm and to protect the nerves and blood-vessels, which lie beneath it, from injurious pressure, when we grasp any substance firmly in the hand. The fan-like expansion of this tendon in the palm is called the "palmar fascia." It is very strong, and is connected, below, with the ends of the metacarpal bones, and with the sheaths of the fingers. The bundle of muscles near D forms what is called the "ball of the thumb," and serves to move the thumb in various directions.

Beneath these three muscles which bend the wrist and strengthen the palm lies another set of muscles (A, B, fig. 66) which bend the thumb and fingers. They pass from the bones of the forearm, and end in long tendons or "leaders" which run over the wrist and palm and along the fingers and are firmly connected with the last phalanges of the fingers. They lie close to the bones in their whole course, and are held in their places by sinewy cross bands and sheaths

which are seen, both at the wrist and in the
fingers, in fig. 65.

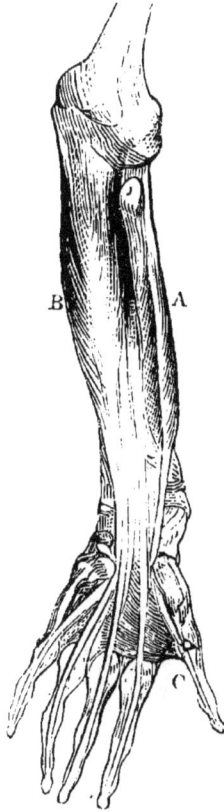

Fig. 66.

Muscles and tendons of hand.

Fig. 67 represents the muscles on the back
of the forearm. The tendons pass from them,
and run, some to the wrist and extend, or bend
backwards, the wrist upon the forearm, some
to the thumb and extend the several joints of

the thumb; and others run to the back of the
fingers. These leaders lie nearer to the skin

Fig. 67.

Muscles and tendons on back of forearm and hand.

than do those on the palmar aspect; and most
of those which go to the thumb and fingers may
be distinguished through the skin. The short
muscles (A, A) situated upon and between the
metacarpal bones pass from them to the sides

of the fingers; some of these serve to spread the fingers out from one another, while others have the effect of drawing them together. There are several such small muscles on both surfaces of the hand, but I must not detain you by a description of them; and there are other little muscles passing from the flexor tendons to the phalanges, which have been called *fidicinales*, from their assisting in performing the short quick motions of the fingers, and from their being, accordingly, called into action in playing upon the violin and other musical instruments.

Movements of the Thumb.

In its adaptation to the purposes of bearing the weight and ministering to the locomotion of the body the human foot excels that of any other animal; and, unquestionably, the human hand is not less preeminently distinguished by the nicety, the variety, and the freedom of its movements. This shown by the manner in which it can be twisted about, by the exquisite play of the fingers; and it is shown, above all, by the mode in which the thumb can be moved to and fro, can be opposed to the other fingers, and to any part of them individually and collectively, and can be folded beneath them or clasped upon them as occasion may require.

The power which the thumb possesses, and gives to the hand, is signified by its name— "pollex,"—derived from the Latin word *pollere,* which means to have power. Some have supposed that the word "poltroon" is derived from *pollice truncato,* and signifies one so cowardly that he has submitted to have his thumb cut off in order that he may be incapacitated for fighting.

The faculty which we possess of moving the thumb in the way I have mentioned, athwart the other fingers, and of touching any part of the palmar surface of either of them depends, partly, upon its being set, not quite in the same plane with them, but, obliquely, so that when it is moved towards the palm it faces or opposes the other fingers; and, partly, upon the nature of the joint between its metacarpal bone and the bone of the wrist with which it is connected.

This joint is so constructed as to admit of three different movements. *First,* the thumb can be moved forwards or backwards, that is, towards, or, away from, the palm. *Secondly,* it can be "adducted" or "abducted," that is, approximated to the forefinger or inclined away from it.

Thirdly, it can be " circumducted," that is, its extremity can be made to describe a circle, as in "twiddling the thumbs." These several movements are effected with great power and rapidity by means of the bundle of muscles which forms the " ball of the thumb" (fig. 65. D), as well as by the long muscles and tendons which descend, from the forearm, to the thumb.

Movements of the metacarpal bones of the Fingers upon the Wrist.

The movements of the thumb, through the medium of its metacarpal bone, upon the wrist are much more free than those of any of the other fingers. The middle finger, indeed, has its metacarpal bone set upon the wrist so fixedly as to admit of scarcely any such movement. The forefinger can be thus moved a little; the ring finger more and the little finger still more.

You may easily prove this by taking the knuckles or heads of the respective metacarpal bones of one hand between the fingers and thumb of the other hand, when you will find that you can press the knuckle of the little finger backwards and forwards in a very perceptible manner. Then try the knuckle of the ring finger; the movement is distinct, though not so free as in the case of the little finger. The knuckle of the forefinger you will find to be almost fixed; and in that of the middle finger you will be unable to perceive any movement at all.

In fact the joints of the matacarpal bones of the fingers with the wrist resemble those of the outer four toes with the tarsus; and the drawings of these joints of the foot (figs. 22 to 25) will serve sufficiently well to illustrate those of the hand.

These movements of the metacarpal, or knuckle, bones upon the wrist enable us to increase or diminish the hollow of the palm by bringing its edges more or less forward. Thus, when we make a cup of the hand we bring forward the metacarpal bones of the thumb and the little finger, wrinkling the skin of the palm; and when we spread the hand open we carry those bones backwards, rendering the skin of the palm tense.

These movements, moreover, enable us to bring the little fingers and the thumb more easily into contact.

Have you ever wondered what advantage is gained by the fingers and thumb all differing from one another in length; or don't you take the trouble to reflect upon little matters of this sort? If you have, I would ask you now to re- mark that there is, in the several fingers, a rela- tion between their shortness, their position near the edge of the hand, and the amount of mobility of their metacarpal bones upon the wrist. Thus the finger which is in the middle of the hand is the longest, and its metacarpal is the most fixed. The fore-finger is not quite so long; and its meta- carpal is rather less immovable. The ring-finger comes next in shortness and in the mobility of its metacarpal. Then the little finger; and the thumb which is much shorter than any other has also its metacarpal much more moveable.

Observe, further, that, when the fingers and thumb are separated from one another, and then bent, the middle knuckle-bone remains stationary, but the others are advanced a little forwards, each to an extent proportionate to its mobility upon the wrist, and to the shortness of the finger. The fore-finger is, by this means, advanced a little, the ring and the little fingers more, and the thumb most of all. And the result is, that the tips of the fingers and the thumb come all to a level, and form, with the palm, a great hollow in which we can grasp any substance, a cricket-ball, for instance, and hold it very firmly. The length of the several fingers .and the thumb is, there-

fore, just so regulated, in relation to their mobility upon the wrist, as to give us this power.

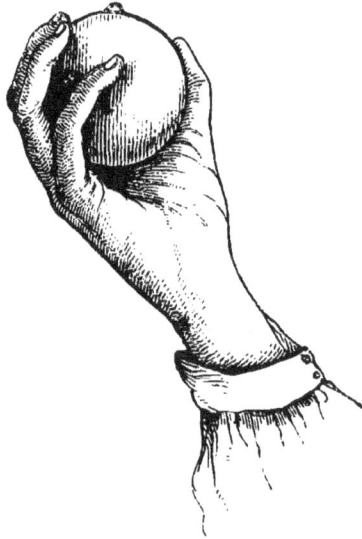

Fig. 68.

You may observe, also, that when the fingers
and the thumb are spread out the space between
the thumb and the fore-finger is considerably
greater than either of the spaces between the other
fingers. Then, by a slight movement, the thumb
takes up a position in front of, or opposite to,
the fingers; and in grasping any substance it
has to antagonise the pressure exerted by all
the fingers. Hence it needs to be much stronger
than they are, and to be wielded by more
numerous and more powerful muscles.

The MIDDLE FINGER is not only the longest and the largest finger; it is also, to a certain extent, the centre about which the others move. Thus, when the fingers are bent down into the palm, their tips all converge towards the middle finger; and when they are spread out, they all diverge from it. Its greater length and the greater prominence of its knuckle, expose this finger to injury more than any of the others; which may account for the fact that Surgeons are called upon to amputate the middle finger more frequently than either of the other fingers or the thumb.

The FORE-FINGER has the greatest range of independent movement. Hence it is used to point with, and is called the "Index" or "Indicator" finger.

Writing.

In WRITING the pen should be held between the pulps of the fore and middle fingers and the thumb, in contact with all three, and firmly lodged between them. The down-stroke is made by bending the phalanges of the fingers and the thumb inwards and the metacarpal bone of the thumb outwards; and the up-stroke is made by straightening all the joints of the fingers and thumb.

The hand rests lightly, not upon its whole edge, but, upon the hindermost and foremost parts of the edge, that is, upon the pisiform bone of the wrist and upon the little finger near the end, so that it can be moved easily along the paper, and can be inclined, or rolled, a little to either side. The obliquity of the stroke is not imposed by mere arbitrary rule, but is in accordance with the direction in which there is the freest movement of the fingers and thumb when thus holding the pen. Make the experiment for yourselves of moving the pen in different directions, and you will soon be satisfied that the writing-master has nature on his side in insisting on a particular movement as well as a particular mode of holding the pen. Some persons make the strokes vertical, or slope them the wrong way; but in either case the writing is stiff and awkward; it is not natural.

The custom of writing from left to right may also be regarded as correct or natural, inasmuch as we can move the pen more freely upwards and *out*wards than upwards and *in*wards. Hence the light *up* or advancing stroke—that which connects a letter with the one which follows it—is most easily made *out*wards or to the right; and the letters are, consequently, made to follow one another in that direction. To understand this more clearly make a down-stroke upon paper in the usual manner; you will then find that you can make up-strokes from any part of it more easily to the right than to the left; and if you make a series of continuous up-and-down-strokes at a little distance from one another, the up-stroke is, not merely habitually, but naturally, made fine and inclined to the right, whereas the down-stroke is made heavier or thicker and is sloped to the left. Moreover, you will perceive that the hand slides along the paper more easily from left to right than from right to left.

It is worthy of remark that the writing of all that great class of languages called Indo-European, which includes Sanscrit, Greek, Latin, and many others, with our own, is from left to right; whereas nearly all the writing of another great class called the Semitic, which includes the Hebrew and Arabic, is in the opposite direction, viz. from right to left. Some nations write in perpendicular columns, the letters being placed under one another, of which the Chinese affords an example. But either of the two latter methods must be inferior to the Indo-European style in ease of execution and expedition.

Reason for the Ring being usually placed upon the Fourth finger.

The RING-FINGER has less independent move-ment than either of the others. It cannot be bent or straightened much without being accompanied by one or both of those next it. This is, partly, because its extensor tendon is connected, by means of a band of fibres, with the tendon on either side of it. You may discern these connecting bands working up and down under the skin of the back of your hand when you move the fingers to and fro (they are represented in fig. 67). The ring-finger is, therefore, always, more or less, protected by the other fingers; and it owes to this circumstance a comparative immunity from injury, as well, pro-bably, as the privilege of being especially selected to bear the ring in matrimony. The left hand is chosen for a similar reason; a ring placed upon it being less likely to be damaged than it would be upon the right hand.

Other reasons have been given for this prefer-ence. It has been attributed to a notion among the ancients that the ring-finger is connected with the heart by means of some particular nerve or vessel, which renders it a more favourable medium

than the other fingers for the reception and trans-
mission of sympathetic impressions; the left hand
being selected, in preference to the right, because
it lies rather nearer to the heart.

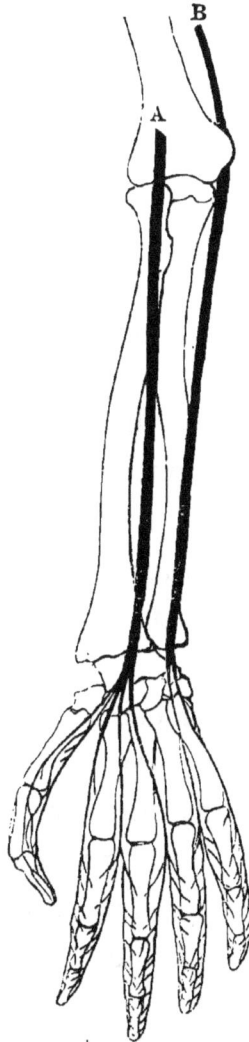

Fig. 69. Nerves of hand.

Some slight foundation for such a notion might be imagined to exist in the fact (supposing the ancients to have been acquainted with it) that the distribution of the nerves to the ring-finger is rather peculiar. The peculiarity will be readily understood by reference to the accompanying drawing (fig. 69). Two chief nerves are seen descending, in their course from the brain, along the arm and forearm, to supply sensation to the palmar surface of the hand. One (A), the larger of the two, passes in front of the middle of the wrist, and divides into branches which are distributed to the skin of the thumb, of the fore and middle fingers, and of the *outer* side of the ring-finger. The other nerve (B) lies on the inner side of the forearm and wrist, and its branches go to the skin of the little finger, and of the *inner* side of the ring-finger. You see, therefore, that there is, in this finger, a meeting of the branches of the two nerves; the two sides of the finger being supplied by different nerves. It would be a mistake, however, to suppose that it gains any superiority in sensitiveness or sympathetic relations by this arrangement; and this distribution of the nerves certainly does not offer so probable an explanation of the selection of that finger for the honourable office of ring-bearer as the one I have suggested.

I must remark, here, that the nerve (B), in

passing from the arm to the forearm, lies on the inner side of the back of the elbow, and is popularly known by the misnomer of the "funny-bone[1]." It lies, pretty much out of harm's way, in a well-protected channel between two bones. Nevertheless, it is now and then hurt; and you know that when the "funny-bone" is struck, a peculiar pain, or tingling, is experienced along the little finger and the adjacent side of the ring-finger.

The practice of wearing rings upon the hand is a very ancient one. In some instances they were badges of slavery. More generally they were marks of high esteem or authority; as when "Pharaoh took off his ring from his hand and put it upon Joseph's hand," and when "Ahasuerus took off his ring, which he had taken from Haman, and gave it to Mordecai." The Roman knights also wore rings of gold. Sometimes rings were worn as charms against diseases; a practice which has been revived in our own day. They were placed upon any of the fingers, and upon the right hand as well as the left. Thus we read in Jeremiah, "though Coniah the son of Jehoiakim king of Judah were the signet upon my right hand." The preference of the left hand and of the ring-finger seems to be comparatively modern, originating, probably, when

[1] It has been suggested, probably by *Punch*, that it is called the "*funny-bone*" because it lies near the "*humerus*."

the ring was made lighter and more fragile, and was, at the same time, adorned with precious stones, and when it became, therefore, desirable to place it upon the part of the hand where it is least exposed to injury.

The Monkey's Hand.

Most of you have spent some time in watching the inmates of that interesting part of a zoological collection, the MONKEYS' cage, and have observed how nearly the hand of that animal resembles the human hand, in the presence of a thumb, in the variety and celerity of its movements, in the facility with which it can catch and pick up objects and hold them up to the mouth, and in some other points. A little closer observation, however, will show that there are some differences between the two. The several parts do not bear the same relation to one another in the Monkey's hand which they do in the human hand; neither have they quite so great variety or range of movement. The hand is altogether narrower, and straighter. The thumb is shorter and less strong, scarcely reaching beyond the knuckle of the fore-finger. The fingers, on the contrary, are longer and of more uniform length; they do not admit of being separated so widely from each other in a fan-like manner; and the metacarpal bones at the edges of the hand, i. e. the metacarpal bones of the thumb and of the ring and little fingers, have not the same amount of play upon the wrist. Hence the thumb and the fingers of the Monkey cannot be opposed to one another so easily as in man; neither can they be so advanced in front of the middle finger as to form a hollow or cup, in the way I described when speaking of the hollow of the palm and the different lengths of the fingers in the Human hand. When you throw a Monkey a nut he usually picks it up and holds it between the thumb and the *side* of the bent fore-finger, not between the tips of the thumb and fingers. The length of the fingers adapts the Monkey's hand well for clasping firmly the branches of trees, and assisting the animal to climb about in its native forests, or to hold on to the bars of its cage; and so the part answers the requirements of the creature better than if these qualities had been sacrificed to a greater regard for variety and range of movement.

The Hand the Organ of the Will.

The human hand is peculiarly an organ devoted to the will, being more directly and completely under its influence than is any other part of the body. The WILL, remember, is that self-directing faculty which can be said to exist, definitely and decidedly, in Man alone, which is associated in him with the responsibility attaching to the selection between good and evil, and which is given to him to fit him to be the reasonable servant of his Maker, and upon which, therefore, his dignity, and his capability for occupying a position between the low animal and the high spiritual world, so much depend. How appropriate is it, then, that the will should have a special organ assigned as its more peculiar minister. It is to the complete subjection of the hand to the will, no less than to the combination of strength with variety and delicacy in its movements, that Man is indebted for his dominion over the rest of the animal world, and for the ability to execute the wonderful works which his genius designs.

When we reflect how essential is the hand to Man's well-being, power, and progress, and upon the infinite variety of purposes which it serves in obedience to the will, we are not surprised that the construction of the foot, indeed of every part of the frame, should have reference to the object of liberating the hand from the subordinate work of locomotion to a degree which we find in no other animal, and of leaving it free to execute its higher offices in a ready and efficient manner.

But, after all, notwithstanding the excellence of its mechanism and its intimate relation to the will, what would the hand be without the reflecting and designing MIND—the mind that can build upon the past and prepare for the future, and so carry on the ever-advancing work of human civilization and progress. Without it Man would remain stationary, like the other animals; and, as age succeeded age, the hand would only suffice to provide the necessary requirements of the body. Nay, even this is saying too much; for without the mind, without, at least, some higher instinctive or reflective faculty than the other animals possess, Man would, in reality, be inferior to them. He would be absolutely unable to maintain his existence, and would be a miserable victim to the fineness of his organisation. His hand would fail to supply him with food, or to defend him against his numerous enemies, or to provide for the protection of his delicate and sensitive frame from the inclemency of the elements.

The real excellence of the human hand—and the remark applies equally to the whole human body—consists, not in the admirable construction of its several parts, nor in their well-adjusted relation to one another, so much as in the adaptation of the whole to the mind that presides over it. This it is that renders Man the lord of the creation, that enables him to subdue all his foes, and gives him, in some measure, power over the elements, so that land and water, fire and air, are made to serve his purpose. By this harmonious co-aptation of mind and body Man is rendered cosmopolitan, being able to thrive in every clime, from the regions of continual snow to those burning equatorial plains where even reptiles perish from the heat and drought, and being able to convert the barren plain into a fertile field, and to draw water out of the stony rock.

At the late meeting of the British Association at Oxford, a gentleman related that he had a monkey which was very partial to oysters, and was very fond of playing with a hammer; but he never could be taught to use the hammer for the purpose of breaking the oyster-shells to gratify his appetite. How wide a gulf does the absence of intelligence in this simple matter indicate between ourselves and the animal that approaches nearest to us!

The Hand an Organ of Expression.

Further, we cannot fail to recognise and admire the adaptation of the hand to the mind at all ages, and under various circumstances; in its weakness and suppleness, and in its purposeless and playful movements in infancy and childhood; in its gradually increasing strength and steadiness as the intellect ripens; in the stiffness and shakiness of declining years; in the iron grasp of the artizan; in the light delicate touch of the lady; in the twirlings, fumblings, and contortions of the idiot; in the stealthy movements of the thief; in the tremulousness of the drunkard; in the open-handedness of the liberal man; and in the close-fistedness of the niggard.

Thus the hand becomes an organ of expression and an index of character. What would the nervous young gentleman in a morning call give to be quit of these tale-telling members; or what would he do without a hat or a stick to employ and amuse them? How effective an auxiliary to the orator is the wave of the hand, or, even, the movement of a finger. Some men, indeed, seem to owe the efficiency of their declamations as much to the hand as to the tongue. I have seen a practised orator (he was a man of the most complete self-possession) quell an excited audience by one determined movement of his hand. It happened to me to hear two of the most celebrated preachers of the day within a short period. In each of them the movements of the hand were remarkable, though very different. In one, the free, impassioned, but natural, and, therefore, easy action of the hand showed a deep and genuine interest in the subject, and helped to waft the fervid sentiments straight from his own heart to the hearts of his audience. In the other, who was a no less accomplished speaker, the constrained and carefully regulated movements of the hands were evidently the result of forethought and study; they were intended to be impressive, but were too obviously done for effect; and, therefore, were far less effective as well as less pleasing.

Our great and venerable orator, as well as high authority on the art of speaking (Lord Brougham), tells us that the subject of a speech should be carefully studied, and the sequences well adjusted. He says that, in the most effective passages, even of practised speakers, the exact words are usually selected beforehand; but he is silent respecting the actions by which they should be accompanied. These, at least, should be unpremeditated; and they will best assist to convey to others the real feelings and emotions when they are the simple result of the natural working of the mind upon the body.

The kind of expression that lies in the hand, being much dependent on the effect of the muscles upon it, is very hard for the artist to catch, though very important to the excellence of the picture. Painters, usually, make the hand a subject of careful study, but rarely succeed in throwing the proper amount, either of animation or of listlessness, into it. In portraits, especially, the hands are a difficult part to treat satisfactorily; yet the artist feels that they are too important not to have a prominent place, and he, commonly, imposes upon himself the task of representing them both in full. I have seen them drawn held up in front, like the paws of a kangaroo, in an otherwise good picture. The stereotyped position in portraits is that one hand lies upon a table, though it, probably, evinces an uneasiness there, while the other rests, perhaps equally uneasily, upon the arm of a chair. Vandyck, in whose paintings the hand usually forms a prominent feature, is considered to have peculiarly excelled in imparting to it a sentimental air imbued with deep pathos.

Shaking Hands.

How much do we learn of a man by his "SHAKE-OF-HAND." Who would expect to get a handsome donation, or a donation at all, from one who puts out two fingers to be shaken and keeps the others bent as upon an "itching palm"? How different is the impression conveyed by the hand which is coldly held out to be shaken and slips away again as soon as decently may be, and the hand which comes boldly and warmly forward and unwillingly relinquishes its hearty grasp? Sometimes one's hand finds itself comfortably enclosed, nursed, as it were, between both hands of a friend, an elderly friend probably; or it is shaken from side to side in a peculiar short brisk manner. In either case we are instinctively convinced that we have to do with a warm and kindly heart. In a momentary squeeze of the hand how much of the heart often oozes through the fingers; and who that ever experienced it has forgotten the feeling conveyed by the eloquent pressure of the hand of a dying friend, when the tongue has ceased to speak?

Why do we shake hands? It is a very old-fashioned way of indicating friendship. Jehu said to Jehonadab, "Is thine heart right as my heart is with thine heart? If it be, give me thine hand." It is not merely an old-fashioned custom; it is a strictly *natural* one, and, as usual in such cases, we may find a physiological reason, if we will only take the pains to search for it. The Animals cultivate friendship by the sense of touch, as well as by the senses of smell, hearing, and sight; and for this purpose they employ the most sensitive parts of their bodies. They rub their noses together, or they lick one another with their tongues. Now, the hand is a part of the human body in which the sense of touch is highly developed; and, after the manner of the animals, we not only like to see and hear our friend (we do not usually smell him, though Isaac, when his eyes were dim, resorted to this sense as a means of recognition), we, also, touch him, and promote the kindly feelings by the contact and reciprocal pressure of the sensitive hands.

Observe, too, how this principle is illustrated by another of our modes of greeting. When we wish to determine whether a substance be perfectly smooth and are not quite satisfied with the information conveyed by the fingers, we apply it to the LIPS and rub it gently upon them. We do so, because we know by experience that the sense of touch is more acutely developed in the lips than in the hands. Accordingly, when we wish to reciprocate the warmer feelings we are not content with the contact of the hands, and we bring the lips into the service. A SHAKE-OF-HANDS suffices for friendship, in undemonstrative England at least ; but a KISS is the token of a more tender affection.

Possibly it occurs to you that the TONGUE is more sensitive than either the hands or the lips. You have observed that it will detect an inequality of surface that escapes them both, and that minute, indeed, is the flaw in a tooth which eludes its searching touch. You are right. The sense of touch is more exquisite in the tongue than in any other part of the body; and to carry out my theory, it may be suggested that the tongue should be used for the purposes of which we are speaking. It is so by some of the lower animals. But, in man, this organ has work enough to do in the cultivation and expression of friendship in its own peculiar way; and there are obvious objections to the employment of it in a more direct manner for this purpose.

The Skin of the Hand.

By the aid of the accompanying drawings you
will be able to form some idea of the structure of
the SKIN of the hand.

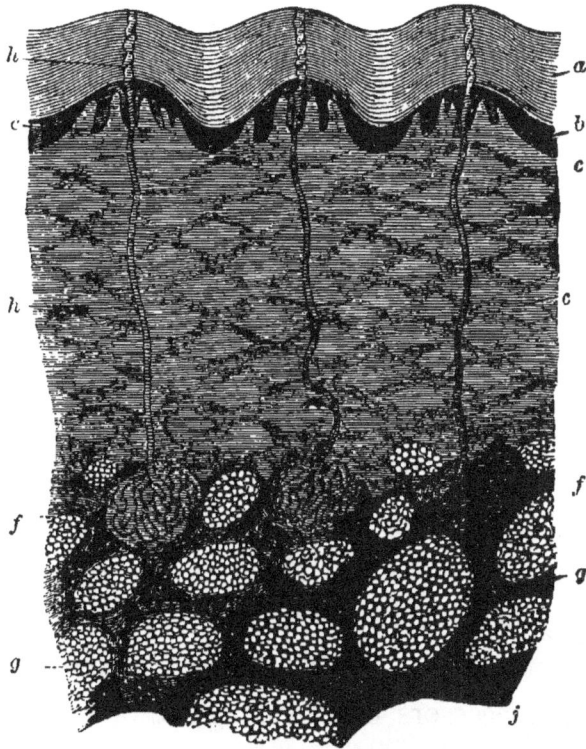

Fig. 70. Skin.

One of them (fig. 70) represents a section of
the skin, made perpendicular to the surface, as

seen under the microscope. It is from the end of
the thumb, and includes three of those delicate
lines, or ridges that are found there.

The superficial, or uppermost strata (*a* and *b*),
are the "Cuticle" or "false skin." The outer layer
(*a*) is hard, horny, and dry. It is composed of
numerous fine scales laid upon one another, like
the tiles upon the roof of a house, but adhering
more closely together, so as to form one continuous
sheet extending all over the body. The outermost
of these scales are continually being shed, peeling
off as scurf, or being rubbed off; and fresh ones are
supplied by the next layer (*b*), which is a softer
material and lies immediately upon the surface of
the "cutis" or "true skin."

This softer layer (*b*) is often called the "*Rete
Mucosum.*" It is made up of minute bags or blad-
ders, named "cells" by anatomists, which grow
and propagate upon the exterior of the true skin,
being nourished by the blood in the skin. Those
which lie nearest the cutis are the youngest and
the softest. Gradually they are pushed outwards
by their successors or offspring; and, as they approach the surface, they become flatter and drier
and more adherent to one another, and are finally
converted into the thin scales of the cuticle. Thus,
there is no real line of division between the cuticle
and the rete mucosum; but the cells of the latter

are gradually transformed into the scales of the former.

The rete mucosum is thicker in the Negro than in the white man, and contributes somewhat to the softness of his skin. It contains also the colouring matter in the form of minute black particles diffused among its cells (fig. 72). These particles disappear, more or less, as the cells become changed into scales; hence the outer part of the cuticle of the Negro is not so dark as the rete mucosum, but, as it is transparent, or nearly so, it allows the dark colour of the rete to show through it.

Persons commonly speak of the cuticle as if it were the whole thickness of the skin. Thus, when a blister has drawn, they say the *skin* is raised; whereas it is only the *cuticle*. This is forced off from the skin by the fluid effused into its softer layer—i. e. into the rete—in consequence of the irritating influence of the blister.

The cuticle has no nerves, and, therefore, no feeling. It may be cut or torn without pain. The snipping of a blister with the scissors is not felt, because the cuticle only is touched. It forms a covering to the whole surface of the body, and is invaluable as a means of preventing too great evaporation. Without it we should be dried up, almost mummified, by the end of a summer's day. It also protects the delicate sensitive skin underneath. How sore is the knuckle when the cuticle has been rubbed off! The cuticle has, moreover, the accommodating property of becoming thickest where it is most wanted, as on the sole of the foot, and on the palms of the hands of blacksmiths, and artizans, and persons who handle the oar. And if any other part of the body be subjected to much friction, for instance, the knees of housemaids, or the shoulders of men who carry packs, the cuticle soon becomes thickened there.

Beneath the cuticle lies the "Cutis" or "True Skin" (c, fig. 70, and c and d, fig. 71). It is a tough structure consisting of interlacing fibrous and fine muscular tissue, and contains the blood-vessels and nerves. The cuticle may be pared off without any bleeding; but directly the skin is wounded the blood flows. The cutis does not present an even surface next the cuticle, but shoots out into a number of little finger-like processes, called "Papillæ," which project into the contiguous soft stratum of the cuticle, and are embedded in it. Thus the superficies of the skin is increased; and as the blood-vessels and nerves of the cutis are continued into the papillæ, they contribute very greatly to the sensitiveness of the skin. They are most numerous in parts where the sensitiveness of the skin is greatest; for instance, they are more numerous on the palmar, than on the dorsal, surface of the hand. Near the ends of the fingers

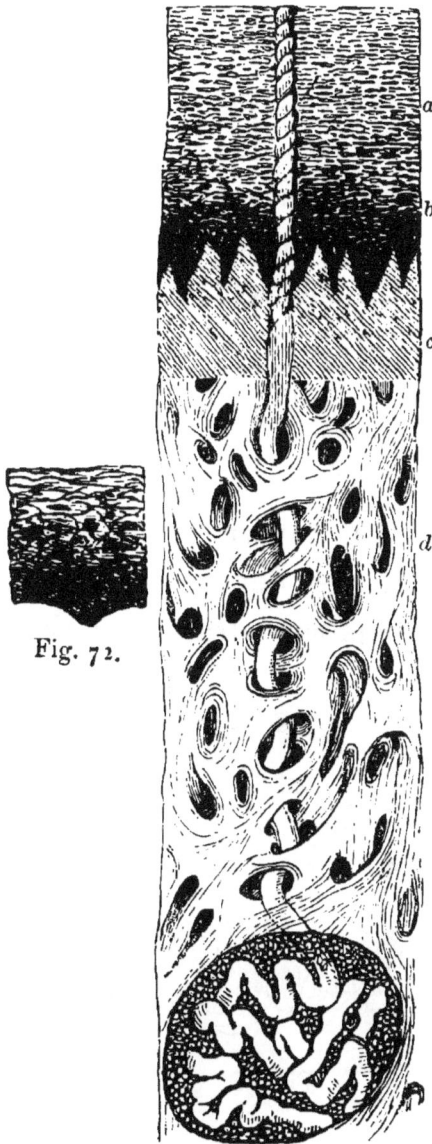

Fig. 72.

Fig. 71. Skin.

and thumb they are arranged in a linear manner,

forming the delicate ridges that encircle the cones
of the pulps. Sections of these ridges are represented in fig. 70.

The superficial or papillary part of the cutis
is of finer and more delicate structure than the
deeper or fibrous layer, and is, therefore, sometimes
described as a separate layer. It is so represented
in the accompanying figure (71, *c*).

As we are upon the subject of the cuticle and
the papillæ, I will take the opportunity to say a
word respecting two diseases of these structures,
in which most of you, probably, have a personal
interest. I mean "Warts" and "Corns."

A WART (fig. 74) depends chiefly on a diseased
state of the papillary stratum of the skin. The

Fig. 73. Corn.

Fig. 74. Wart.

papillæ become coarse and grow up beyond the
level of the surrounding skin, so as to present an
uneven or "warty" surface. They carry a layer of

cuticle before them. This layer is usually thin, so that the wart bleeds easily when it is rubbed. Sometimes, however, it is very thick and hard like a piece of horn. We, now and then, hear of a horn growing upon some part of the body, perhaps on the forehead. Such a horn is, usually, nothing more than a conical mass of cuticle formed upon the surface of a large wart. Warts are generally caused by something irritating the skin, as dirt or soot rubbed into the cuticle. For this reason they are more frequent upon the hands than upon other parts of the body.

In a CORN (fig. 73), also, the papillæ are somewhat enlarged; and this accounts in part for the great tenderness of corns. But the primary and essential feature of a corn is a thickened state of the cuticle. This is caused by too great rapidity in its formation, and is, usually, dependent upon pressure, especially if the pressure be combined with some friction. Hence corns are most commonly found upon the foot, and upon the parts of the foot, where the skin is subject to pressure and rubbing against the shoe. The drawing shows the appearance presented by a vertical section through a corn and through a small portion of the skin on either side. The accumulated layers of cuticle are seen, and the enlarged papillæ shooting up into them. I need scarcely add that it is owing to ignorance, or something worse, when corn-cutters talk of curing the malady by taking out the *roots;* for, corns, evidently, have no roots.

One word of advice about corn-cutting. Most persons have some experience in this art, and some opportunity of practising it on themselves; and many pride themselves on their skill in it. The usual plan is to shave off layer after layer from the whole surface of the corn; and this, by lessening the projection of the corn, may give relief for a few days, though it does not always do that. Soon, however, the distress returns; and the area of the corn increases after each operation. Now, I would have you observe that it is at the *middle* of the corn that the papillæ are most enlarged; and it is here that the formation of cuticle goes on most quickly, giving rise to the little white cone or cones often seen in a corn, and sometimes wrongly called the roots. The proper mode is to confine the cutting to this part, and to remove as much of the thickened cuticle as you can from this spot, digging, as it were, a hole in the middle and leaving the circumference intact. The circumference, which is not usually tender, thus forms a wall round the excavated centre and defends it from pressure; and great relief is experienced. Further benefit will be found from covering the corn with some soft

adhesive plaster; and you may sometimes, with
advantage, lightly apply common caustic before
putting on the plaster. If you follow these directions carefully you may be your own chiropo-
dists, and almost defy your bootmakers.

If, in cutting a corn, you go too deeply, you
will wound the tops of the papillæ and cause some
bleeding; this is not however usually followed
by any ill consequences.

Nails.

Almost all vegetable as well as animal surfaces are covered with some kind of cuticle. It
forms the smooth exterior of a leaf and the
rind of an apple; and the soft down of a moth
or a butterfly, the scales of fish, the feathers
and claws of birds, the quills of the porcupine,
the horns of oxen and the hoofs of the horse
are examples of modifications of cuticle. NAILS
and HAIR are also of this nature. They are both
continuous with the cuticle, and peel off with it
when it is, by any process, separated from the
skin. Both are formed, like the cuticle, of compressed plates or scales matted together; and
these are continually being shed or rubbed off
on the one side, and supplied from the rete mucosum on the other.

The rete mucosum, it should be stated, extends over the whole surface of the body. In most situations, as already mentioned, it is the medium from which the ordinary cuticle is produced; but on the upper part of the ends of the fingers and toes it is converted into nail, and in the hair follicles, as I will presently describe, it is transformed into hair.

The drawings will help you to understand the relation of the nails to the cuticle and the cutis. In the upper of the three (fig. 75) the nail with

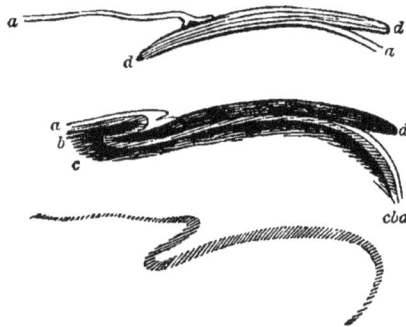

Figs. 75, 76, 77. Longitudinal sections of Nail.

the cuticle has been detached from the cutis, so that the continuity of the two, at either end, is shown. In the middle one (fig. 76) it is represented lying in its bed in the cutis; its thin hinder edge being received into a furrow made for it in the cutis. The layer of rete mucosum (*b*) extends behind and beneath it, between it (*d*)

and the cutis (*c*), and continually adds fresh
material to the nail, just as, in other parts, it
adds to the substance of the cuticle. The cuticle,
or white line (*a*) is continuous with the nail at
the sides as well as at either end. The lower
figure (77) shows the bed of the cutis in which
the nail reposes, the nail as well as the adjacent
cuticle and the rete having been cleared away.

Thus the addition from the rete—in other
words the growth of the nail—takes place at
the hinder edge and at the under surface. In
consequence of the addition from *behind* the nail
is increased in length and is pushed forward ;
and as it advances forwards it receives accessions
from *beneath*, which increase its thickness and
strength. Unless they be cut, or worn down,
the nails grow to an indefinite length ; and,
when they extend beyond the tips of the fingers,
their edges are bent in towards each other, and
they become curved like claws. This tendency to
a convex form is shown also if the nails be not
properly supported by the pulps of the fingers.
For instance, when persons become emaciated
the pulps of the fingers usually participate in
the general wasting and the nails become curved.
Hence this shape of the nails has been regarded
as an indication of consumption. You will understand, however, from what I have said that it

is not really a symptom of any one particular disease. It simply indicates that, from some cause or other, the nutrition of the body is not properly maintained.

The Dervishes in some parts of Asia allow the thumb-nail to grow long, and then pare it to a point, so as to be able to write with it. Dr Wolff, the Eastern traveller, has told me that he has repeatedly seen this done, and that he has in his possession manuscripts written in this way.

Beneath the nail the cutis is disposed in a series of parallel ridges (fig. 78) with intervening furrows. These take the same direction as the

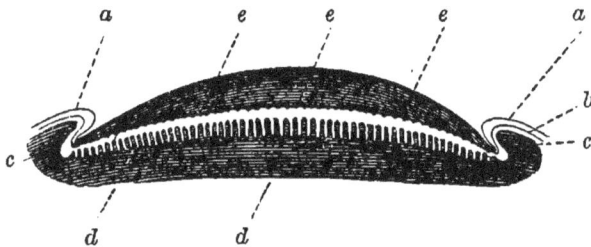

Fig. 78.   Transverse section of Nail Rete and Cutis.

nail, and, indeed, give rise to the fine lines that you see upon the surface of the nail. The rete mucosum, or deep soft layer of the nail, extends into the furrows between the ridges, just as the soft stratum of the cuticle extends between the papillæ of the cutis. It thus serves to keep the nail steady in its place, while it permits a certain amount of movement, and allows the nail to slide forwards upon the skin under the pressure caused by the growth at its hinder edge.

A little in front of the root of the nail the ridges of the cutis suddenly become larger and more vascular. This gives a pink hue to the nail in the greater part of its extent; while the hinder portion, separated from the front by a crescentic line, is white, in consequence of the subjacent cutis being there, more pale. You will, at once, recognise the distinction between these parts by looking at your own nails.

The ridges and furrows serve, like the papillæ in other parts of the skin, to increase the surface of the cutis; and, by affording more space for the distribution of the vessels and nerves, they contribute to the sensitiveness of the part, and account for the severe pain which is caused when any foreign body is thrust under the nail. The pulp in the interior of a tooth, and the frog of a horse's foot, are also instances in which an exquisitely sensitive structure is placed beneath a hard or horny substance. The object, in each case, is the same, viz. to give the power of taking cognizance of impressions which are made upon the surface.

Hairs.

HAIRS may also be regarded as modifications
of the cuticle, because, like the nails, they are continuous with the cuticle, and are formed from
the rete mucosum. Each hair (figs. 79 and 83) is
received into a depression of the cutis, which is
called a "follicle," and which is lined, as far as the
bottom, by cuticle (*a*), and rete mucosum (*b*). At
the bottom of the follicle (*d*) the cuticle is absent,

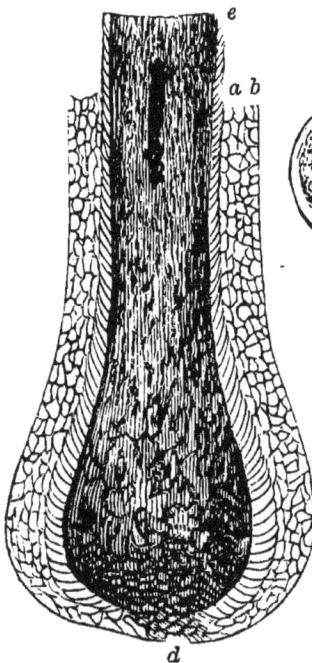

Fig. 79.   Hair.

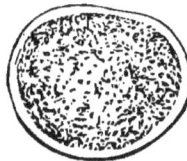

Fig. 81.

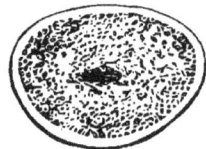

Fig. 82.

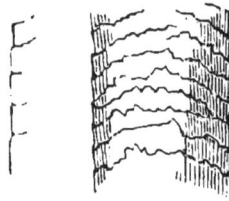

Fig. 80.

and the hair rests, directly, upon the rete; and, at
this part, the rete, instead of being converted into

cuticle, as it is at the sides of the follicle, becomes transformed into hair, in the following way.

The cells of which the rete is composed swell out as they ascend, and so form the soft "bulb" of the hair. The outermost cells are gradually flattened, and assume an imbricated arrangement, overlapping one another like the tiles upon a roof (fig. 79, *e*, and fig. 80); and those in the interior are elongated, so as to be converted into more or less distinct fibres. The cells nearest the middle, or axis, of the hair remain moister and softer than those nearer the exterior, and form what is sometimes called the "marrow" of the hair (figs. 81 and 82).

The colour of the hair is given by the presence of minute grains of colouring matter, like those in the cuticle of the Negro. They are formed in the cells at the root of the hair, and pass up with them into its structure. The quantity of colouring matter is usually slight in infancy and childhood, and increases during adolescence. Hence the hair becomes darker as we grow up. It is more or less deficient in the grey hair of old age; and in the instance of Marie Antoinette, and others whose hair is said to have turned grey in a few hours, the colouring matter is supposed to have been destroyed by some fluid, formed from the blood, and passing, through the pulp, into the hair.

The hairs serve to protect the skin; and, as a general rule, they are most abundant upon the parts which are most exposed, and which, therefore, stand most in need of such protection. They are scattered over the back of the hand. On the palmar surface they are not required, and they would have interfered with the sense of touch; and we do not, accordingly, find them there, nor upon the sole of the foot, nor upon the edges of the lips. In certain parts of some animals, however, they serve as valuable adjuncts to the tactile organs by extending the range within which the contact of surrounding substances is felt. Thus the whiskers of the cat are set upon papillæ so sensitive that the slightest touch upon any part of the hair is felt; and the animal is thereby assisted in threading its way in the dark. This provision, added to the mode in which their feet are muffled with soft hair and their claws are retracted, enables the members of the feline tribe to steal with almost absolute stillness upon their prey.

Oil-glands.

There are also in the skin a number of little GLANDS. One set of these are called "oil-glands;" for their office is to furnish an oily, or waxy, sub-

stance, which serves to keep the skin soft and pliable, and defends it against too much moisture, or
too great dryness of the atmosphere. They are
usually, as shown in the accompanying sketch, (fig.
83, *g*, *g*) connected with the hairs, lying beside them;
and their ducts—the little tubes that carry off the
oily matter formed in them—open either into the
hair follicles, or penetrate the cuticle at some other
part. They are not found on the palms of the hand
or the soles of the feet, because those parts are, in
great measure, sheltered from atmospheric influ-

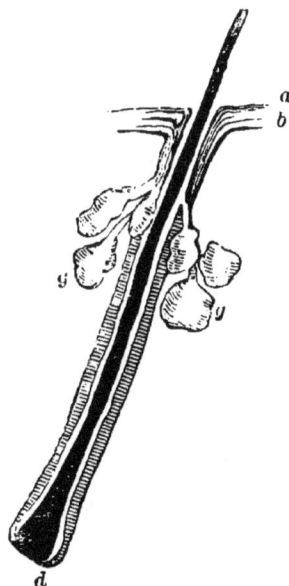

Fig. 83.   Hair, and Oil-glands.

ences, and are well moistened with perspiration.

When the dry easterly winds prevail one is disposed to wish that these glands were more numerous on the back of the hands; for a more liberal supply of their secretion would, probably, prevent the disagreeable chapping to which we are subject at those times. As a substitute we resort to some unctuous matter, such as glycerine, which if frequently applied in small quantities performs, to some extent, the part of the natural secretion in keeping the cuticle soft and supple, and so preventing its cracking.

The secretion of these glands has an odour, the purpose of which, in man, is not very obvious. It is faintest in the highest and most civilized nations. In none is it very agreeable; and persons are fain to conceal it by substituting some other odour, as that of lavender or eau-de-cologne. Unfortunately the choice is not always so refined; and one is, sometimes, disposed to think that the natural odour must be very bad, if the substitute be preferable. The odour varies at different parts of the body; it varies also in different persons, sufficiently to enable the acute nose of the dog to track one particular man among a thousand.

Sweat-glands.

To revert to the figure (70) at page 165, the little masses at *g, g,* are grains of fat lying in the meshes of the deeper strata of the skin, or in the structure just below it. And the little balls of twisted tube (*f, f*) are GLANDS that secrete the PERSPIRATION; for, the perspiration does not ooze up from the whole surface of the skin, but has a regular system of factories for its formation. A fine tube (*h*) is seen passing from each of these "sweat-glands," as they are called. It curls in a spiral manner, like a cork-screw, where it traverses the cuticle to open at the surface. On the palmar aspect of the hand most of these tubes or ducts open along the tops of the fine ridges which are there seen; and with a magnifying glass of moderate power you can distinguish their orifices on the flattened tops of the ridges on your own fingers. These are the "pores of the skin," respecting which we hear so much, and through which the Roman Bath brings such streams from the subjacent glands.

The sweat-glands are scattered all over the body, but are especially numerous in the palm and in the sole; and the moisture issuing from

them tends to keep the skin of these parts soft
and moist, and so fitted for the reception of tactile
impressions. The quantity of fluid furnished by
them varies a good deal in different persons, and
under different circumstances. In some persons it
is habitually slight; and the hands feel dry and
harsh. Or, what is equally disagreeable, it is su-
perabundant; and the hands are habitually damp,
perhaps, cold and clammy, staining the gloves and
soiling everything they touch. In fever the perspiration is defective; and the dryness and heat of
the palm are often the first symptoms of an acces-
sion of fever that attract the patient's notice.

We all know that perspiration is usually increased by exercise, or by the application of warmth
to the surface, as by the hot air in the sudatorium
of the Roman Bath; and then, by its evaporation, it
cools and relieves the body, and contributes to our
comfort. We know, too, that it is liable to be
increased by any thing that produces a depressing
effect, and that it then induces an uncomfortable
sensation, chilling the surface too much, and making it cold and clammy. Most of you have experienced the discomfort of the cold sweat caused by
fright; and some of you may have felt the cold,
clammy hand of one who was suffering under the
shock of a severe accident or the prostration caused
by the sudden onset of a dangerous malady. Why

perspiration should occur under these very different conditions, producing, at one time, so much
relief, and, at another, so much additional discomfort, it is not easy to say.

The sense of Feeling and of Touch in the Hand.

I have mentioned three parts of the body as
remarkable for the acuteness of the sense of touch,
namely, the TONGUE, the LIPS, and the HANDS.
Now in each of these the skin is richly supplied
with nerves and blood-vessels; and it is also thick
and lies upon a soft cushiony substance, so as to
be yielding and to admit of being applied accu-
rately over any surface with which it is placed in
contact, and of again resuming its shape when the
pressure is removed.   For instance, the tongue is
so soft and yielding that, when it is applied to a
tooth, it dips down between the inequalities and
gives accurate information of the condition of the
whole surface.   The same is the case with the
edges of the lips, though not in so marked a degree as in the tongue; and each of these parts is
indebted for its great sensitiveness very much to
the delicate soft supple nature of its structure.
The palmar surface of the hand too, though, like
the skin of the sole, it is strong and tough, so as
to offer considerable resistance to injury and to

prove no dainty morsel even to dogs, as we sur-
mise from the narrative of the death of Jezebel, is
yet very soft and yielding. It is also underlaid by
a stratum of fat interwoven with strong fibres of
tissue, just in the same manner as the skin of the
sole of the foot (fig. 46, p. 99).

An accumulation of this fat and fibrous tissue
under the skin forms the "Pulps" at the ends of
the fingers. The slightly conical form and exquisite softness of the Pulps adapts them well for the
examination of the surfaces of bodies; and the
sense of touch is more acute in them than in other
parts of the hand.

In connection with them it is interesting to
observe that the last bone of each finger and of

Fig. 84.　Bones of Finger.

the thumb swells out, at the end, into a nodulated
lump, which serves the purpose both of supporting
the pulp and of giving breadth to the nail. It

also, like the corresponding part of the toe (page 99), affords a basis of attachment for the fibres that run, from the bone, through the pad of fat, to the skin, and give firmness and consistence to the part. The bulbous enlargement at the ends of the phalanges of the fingers and toes is peculiar, or almost peculiar, to man. In most Animals these bones taper to a point; in many they are also curved. Hence the nails are, in them, comparatively unsupported, and they become bent in at the sides and curved in their length, that is to say, they are formed into claws. This is the case, to a considerable extent, in the Monkey. The terminal phalanges of the monkey's digits are more tapering than in man; the nails are more claw-like; and the pulps are less well-formed. This constitutes a not unimportant feature of difference between the hand of that animal and the human hand, in addition to those I have already mentioned.

You have experienced the sensitiveness to cold of the pulps of the fingers and toes; and have, probably, remarked that it is more difficult to keep them warm than any part of the body. I may add that, notwithstanding the liberal allowance of the means of supporting life (that is, blood and nervous influence) which they enjoy, they are very liable to mortify from frost-bite and other causes. I have repeatedly known that to happen when all the rest of the hand has escaped. This must be attributed, perhaps entirely, to their exposed position as terminal parts ; and they share their susceptibility to cold with other parts similarly circumstanced, such as the nose, the elbows, the knees and the buttocks.

It is necessary to make a distinction between the SENSE OF TOUCH and common FEELING or sensitiveness to pain; for they are not quite the same. They are, it is true, very nearly alike, so nearly that we may consider them to be modifications of one another; and it is probable that the same nerves minister to both. Still there is a difference. The sense of touch is the sense of contact with *external* bodies, and enables us to take cognisance of their presence and inform ourselves of their shape, consistence, smoothness or roughness, &c.; whereas common sensation, or the sense of feeling, has an *internal* relation. It imparts to us information respecting the condition of our own bodies or any part of them. By the sense of touch in the tongue, for instance, we judge of the size and hardness of the morsel in the mouth ; and by common sensation we learn that the organ is being bruised or scratched by it. Sensation of pain commonly destroys the sense of touch. Put your finger into a vice, and you may feel both sides of it. Screw it up, and you have nothing but the sensation of pain. If you were to awake in this state you would not, from the mere sensation, know that you were *touching* anything.

As a general rule there is a relation between the degree in which sensation and the sense of touch are manifested in different parts of the body. For instance, I have just been remarking on the acuteness of the sense of touch in the Tongue; and we know that this part is very sensitive to pain. The pain caused by a bite of the tongue is horrible; and so effectually does it serve the good end of warning the tongue to keep within its proper bounds, that that organ very rarely suffers from the pressure of the teeth.

But, forasmuch as sensitiveness to pain serves a different purpose from the sense of touch, namely, as in the instance of the tongue just mentioned, it renders parts alive to injurious impressions, and gives them warning to escape or protect themselves; so it is, as we might expect, most manifested in those surfaces where a slight amount of injury would prove most detrimental.

Thus, the membrane (the conjunctiva) which lines the eyelids and covers the front of the eyeball is exquisitely sensitive to pain. We are reminded of this when anything touches the eye, or when a fly has lodged itself under the eyelid. And, when an operator wishes to ascertain whether his patient is sufficiently under the influence of chloroform he separates the eyelids and puts his finger gently upon the eye, knowing that if no indication be given, by flinching, that the impression made here is felt, it is probable that the patient will not be conscious of the more severe impression to be made by the knife elsewhere. Yet, this membrane is by no means pre-eminently endued with the sense of touch. Indeed, the very acuteness of its sensitiveness to pain quite unfits it for distinguishing the quality of the impressions made upon it. We know very quickly that something is in contact with the eye, but can form no idea what kind of substance it is, whether it be hard or soft, rough or smooth.

In the hand, on the contrary, the sensitiveness to pain, though considerable, is not proportionate to the acuteness of the sense of touch. The sting of the rod on the palm, if my recollection serves me right, is not so sharply felt as it is upon that other region which shares with the hand the privilege of receiving the wrathful attentions of the master; and, yet, that other region is by no means distinguished for acuteness in the sense of touch.

The mode in which sensitiveness to touch and

to pain are adjusted in the hand and in the eye in relation to the functions of these two organs is one of the admirable features of their construction. Suppose the disposition to have been reversed— suppose the hand to have been as tender as the eye—of what use would it have been? The contact of a particle of dust would have caused agony; or, had the eye been no more sentient than the hand, it would soon have been destroyed by the chafing of foreign bodies upon its delicate surface.

How important is the sense of Feeling! more important than any of the other senses; more so than all the others taken together. It is almost universal in the animal kingdom. Indeed, we can scarcely conceive animal existence without it, and are slow to admit that to be an animal which shows no sign of it. Several of the lower animals seem to be destitute of any of the other senses. The POLYPS, for instance, have no sight, hearing, taste, or smell, and are dependent, therefore, entirely, upon feeling for their communication with the external world; and the range of this sense is extended in them by means of their "tentacles" or "feelers" which wave about in the water, and, when they come in contact with foreign bodies, close upon them and draw them towards the oral opening. Thus, the tentacle of the polyp is a sort of rudimentary hand, and, by the aid of feeling, fulfils one important function of the hand, viz. that of the supplying the mouth with food. The sprawling movements of an infant's hands and the tendency which they have to close upon anything—dress, blanket, or whatever it be—and draw it to the mouth remind one forcibly of the feelers of a polyp.

In most of the lower animals, however, the sense of feeling, though present, serving for protection and giving notice of injury, is not very acute. It is not much employed by them for the purpose of obtaining information respecting external objects; and they can scarcely be said to enjoy that modification of it which we call the sense of touch in any high degree. Indeed, the skins of animals have, commonly, such a covering of thick, horny cuticle, scales, feathers, or hair, as is incompatible with a fine discriminating sense of touch.

In many of them, however, some other sense is highly developed. The VULTURE is guided by the smell of carrion for miles and miles; and the dog will, by the same sense, track game where man cannot detect the trace of an odour. Some birds can distinguish objects which are quite out of the range of our sight. The EAGLE, for instance, soars aloft, till it dwindles to a mere speck or is lost to our view, and, then, from that great height, will pounce, with unerring certainty, on an unhappy grouse upon the ground. The sense of hearing is a great means of protection to animals, and necessitates extreme stillness and caution on the part of their pursuers. The DEER, when feeding, directs his eyes upon the ground, and depends for safety, chiefly, upon his hearing, which is so acute that the huntsman is obliged to approach with all possible wariness.

In each of these instances, it may be observed, the acuteness of the particular sense is manifested chiefly in the power it gives to the animal of distinguishing objects *at a distance*. Whereas, in the ability to use the several senses for the nice discernment of the *qualities* of substances and to derive enjoyment from them, man stands quite unrivalled. He alone appreciates the perfume of a bouquet, or takes cognisance of the various shades of colour and of the notes of music; and the sense of touch, which is of especial service in aiding us to an accurate knowledge of bodies, is much more highly developed in man than in other animals.

Fine as the sense of touch usually is in the human hand, it becomes far more so when an unusual demand is made upon it in consequence of a deficiency, or absence, of other senses. The rapidity with which blind persons can read with their fingers is truly astonishing. Some are said to be able to distinguish colours by the feel. (It should rather be said that they are capable of recognising the nice differences in certain substances by which colours are caused; for one can scarcely conceive it possible to distinguish by feeling the colours in a ray of light separated by a prism.) I am acquainted with a lady who has been, not only blind, but deaf and dumb from infancy. The sense of touch is, therefore, almost her only avenue for impressions from without; and it is surprising how much information is conveyed through it, and how quickly. It enables her to hold converse with her relatives, by the language of the fingers, almost as freely and as briskly as others do with the tongue. A few touches are sufficient to transmit a series of thoughts. After one shake of the hand her friends told me that she would recognise me again; and, true enough, although several days elapsed before I again saw her, she made the sign for my name as soon as she touched my hand. At our next meeting I presented my left hand, but was, again, immediately recognised.

Persons who have had much experience in the instruction of the deaf and dumb find that the hand, by means of writing and "dactylogy", or the language of finger-signs, is abundantly sufficient for all the intercourse to which a deaf-mute is equal; and they are, therefore, disposed to discourage the teaching of articulation. Dr Kitto, in his little book " On the Lost Senses," which acquires so much interest from the fact of his being himself deaf and dumb in consequence of an accident, relates that, after he had, with great difficulty, reacquired considerable facility of speech, he found it stood him in little stead. So efficient a means of intercourse had the hand become that, he tells us, he had not occasion for the use of his tongue ten times in a year.

Not only may the hand thus serve, to some extent, as a substitute for some of the other senses; it is also a most important auxiliary to them. Particularly is it so to the sense of sight, by proving, or correcting, the impressions which we receive through the eye. Without its aid we should often fail to distinguish between a real object and a picture or a reflection in a mirror, and should have difficulty in judging of size, shape, distance, &c.

Relation of the Hand to the Eye and the Mouth.

You cannot have watched a game of cricket without being struck by the manner in which the hand acts in harmony with the eye. With what almost lightning-like rapidity it is in the exact place to catch the ball; and with what precision the practised cricketer can throw the ball to a great distance. In this, however, he is surpassed by the wonderful skill with which the Indian throws the lasso. Again, it is enough for the sportsman merely to get sight of the bird; he is scarcely conscious of the process by which the hand directs the gun and pulls the trigger at the exact moment. Still more remarkable is the successful aim when taken, as it occasionally is, without bringing the gun to the shoulder.

In estimating the importance of the hand, you must not forget that the mouth is quite dependent upon it for supplies. In most other animals the jaws are prolonged, forwards, from the cranium, and the head is placed in such a position that the mouth becomes an organ of prehension, and is enabled to provide for itself. But, in man, the head is carried so high above the ground, and the jaws are so shortened and compressed beneath the forehead, that the mouth is of little use in obtaining food. Its abilities and duties are restricted to receiving, masticating, and swallowing; and, if it had to rely upon its own efforts for supplies of food, it would, indeed, be in a poor case. When we look at one of the Sphinxes from Egypt, or at one of the stately Bulls from Nineveh, in which wisdom and power are represented by joining a human head to the trunk and limbs of an animal, the question suggests itself, "How is that mouth to be fed?" In the Centaur and Mermaid this difficulty is overcome by adding the hands, as well as the human head, to the trunk and locomotory organs of the horse in the one instance, and the fish in the other; so that monstrosity does not preclude the means of sustentation. Sufficient incongruities, however, still remain to justify the exclamation

"Spectatum admissi risum teneatis amici?"

In the ELEPHANT the mouth is circumstanced, somewhat, as in man; and the office of feeder is performed by the elongated snout or proboscis. This organ, with its finger-like extremity, is so sensitive and mobile as to be able to pick up small bodies—pins or needles—from the ground, and so strong as to pull down large branches of trees, and gather the fruit from them. It is interesting, in connection with the relation of the hand to the will and the intellectual endowments, to remark that this proboscidean substitute, which fulfils so many of the purposes of the hand, is furnished to the "half reasoning" elephant. The natural sagacity and teachableness of this creature, of which such interesting evidence is given in Sir Emerson Tennent's book on Ceylon, seem to render it quite worthy of the privilege of having an especial organ provided to minister to its will.

Cheiromancy.

The BEAUTY of the hand does not come in for
quite so great a share of admiration as that of the
foot.   Perhaps, because we are less often gratified
with the view of the latter.   Perhaps, because we
are conscious that the foot is even more decidedly
characteristic of the human form than is the hand;
inasmuch as the hand of the monkey approaches
more nearly to the human hand than does the
foot of any animal to the human foot.   Still, we
are by no means insensible to the charms of a
pretty hand; and we prefer that the glove which
envelopes it should be of a material as thin and
pliable as kid, so that it may adapt itself accu-
rately to the part, and not conceal its form.   A
small and delicate hand is thought to be one of
the best signs of high-breeding.   Thus, Byron, who
was no bad judge of such matters, writes

"Even to the delicacy of her hand
There was resemblance such as true blood bears,"

and again,

"Though on more thorough-bred or fairer fingers
No lips ere left their transitory trace."

The LINES upon the palm, or creases formed

in closing the hand, differ a little in different persons. In former times, when men were addicted to the arts of divination, and thought more about the connection between the physical world and the world of spirits, and strove, by a close observation of the former, to penetrate the mysteries of the latter, much attention was paid to these lines. They were named with the names of the Planets and the signs of the Zodiac; and a science grew up akin to Astrology and Physiognomy. CHEIROMANCY was the name given to it; and numerous and voluminous treatises were written upon it. We are told that Homer was the author of a complete essay upon the lines of the hand. That something of the kind was practised among the Romans we learn from a passage in Juvenal, translated, somewhat freely, by Dryden, as follows:

"The middle sort, who have not much to spare,
To cheiromancer's cheaper art repair,
Who claps the pretty palm to make the lines more fair."

You will estimate the value of the science of Cheiromancy when you hear that equal furrows upon the lower joint of the thumb argue riches and possessions; but a line surrounding the middle joint portends hanging. The nails, also, came in for their share of attention: and we are informed that, when short, they imply goodness; when long and narrow, steadiness but dulness; when curved, rapacity. Black spots upon them are unlucky; white are fortunate. Even at the present day Gipsies practise the art when they can find sufficient credulity to encourage them.

Whether any fancy of the like kind gave origin to the notion still prevalent that a wound or injury between the thumb and the fore-finger is peculiarly likely to be followed by Lock-jaw, or whether the notion was grounded on some notable instance in which that fearful malady did actually supervene upon a wound in the situation mentioned, I cannot tell. You may, however, rest assured, that it is quite a fallacy. Lock-jaw may result from a wound in any part of the body, or it may occur without a wound; it is very capricious in its attack; the surgeon does not know when to look for it; it often shows itself when he least expects it; but it is not more likely to follow a wound between the thumb and the fore-finger than a wound elsewhere. I think it well to mention this, because I have often known persons greatly alarmed when they have accidentally cut themselves in the dreaded spot.

Cause of the preferential use of the Right Hand.

Why is man usually RIGHT-HANDED? Many attempts have been made to answer this question; but it has never been done quite satisfactorily; and I do not think that a clear and distinct explanation of the fact can be given.

There is no anatomical reason for it with which we are acquainted. The only peculiarity that we can discern is a slight difference in the disposition, within the chest, between the bloodvessels which supply the right and the left arms. This, however, is quite insufficient to account for the disparity between the two limbs. Moreover, the same disposition is observed in left-handed persons, and in some of the lower animals; and in none of the latter is there that difference between the two limbs which is so general among men.

Is the superiority of the right hand real and natural, that is, congenital? or is it merely acquired? I incline much to the latter view; because all men are not right-handed; some are left-handed; some are ambidextrous; and in all persons, I believe, the left hand may be trained to as great expertness and strength as the right[1]. It is so in those who have been deprived of their right hand in early life; and most persons can do certain things with the left hand better than with the right.

Nevertheless, though I think the superiority of the right hand is acquired and is a result of its more frequent use, the tendency to use it, in preference to the left, is so universal that it would seem to be natural. I am driven, therefore, to the rather nice distinction, that, though the superiority is acquired, the tendency to acquire the superiority is natural.

It may be argued that the tendency must be based upon something physical, and that, therefore, a tendency to superiority implies an actual superiority. This may be so; but I do not think that we are quite in a position to assert that it is so. We perceive that there is a tendency to the preferential use of the right hand; but we do not know upon what that tendency depends, and have, therefore no right to assert that the cause of it lies in the construction of the limb or of the

---

[1] In the tribe of Benjamin "there were seven hundred chosen men left-handed; every one could sling stones at an hair breadth, and not miss." Judges xx. 16. When David was at Ziklag there came to him a company of men who "were armed with bows and could use both the right hand and the left in hurling stones and shooting arrows out of a bow." 1 Chronicles xii. 2.

parts which supply the limb with blood and nervous influence, or, indeed, upon any strictly physical cause whatever.

It may be a tendency like that of certain animals to make their holes and nests in particular places and in particular ways, to watch for their prey at particular spots, to migrate in certain directions at particular periods, and to group themselves in a particular order during their travels. Such tendencies, or "Instincts" as they are often called, may possibly be the result of a peculiar conformation of the several animals; but it is, at present, by no means certain that they are so.

I have said that man is the only animal in whom a preference in the use of the limb or limbs of one side is shown. This is a consequence of the fact that he is the only animal who has occasion to use the limbs of the two sides separately, or who is in the habit of doing so. Even in the rudest state of society this habit is engendered in him from a very early period, as in carrying a stick, throwing a spear, and in a variety of ways. The habit increases as he becomes more civilized, owing to the greater number of offices which the hands are called upon to perform; and the necessity for using the hands separately would, of itself, lead each individual

to the employment of one more frequently than
the other; but that that one should so univer-
sally be the right hand, seems to be accounted
for only by reference to some natural tendency.
The imitative propensity in man and the convenience of uniformity of modes of action are scarcely
sufficient to account for it.

I will not detain you by dwelling upon the
effect which the superiority of the right hand
has in giving a slight superiority to the right leg
and the right eye, and will content myself with
mentioning a single beneficial result of the preferential use of one hand, viz. that by it, we
acquire a greater degree of skilfulness and dex-
terity than we should do if both hands were
equally employed. The exclusive use, for instance, of the right hand in writing, cutting, &c.
gives it a greater expertness than either hand
would have had if both of them had been accustomed to perform these offices. Hence, we usually
find that persons who are left-handed are rather
clumsy-fingered, because, although, in them, the
left hand is used for many purposes which are
commonly assigned to the right, yet the conven-
tionalites of life interfere a good deal. The pen
and the knife, for instance, are still wielded by
the right hand. Accordingly such persons are
neither truly right-handed nor truly left-handed;

and they do not commonly acquire so great skill in the use of either hand as do those whose natural tendency is more in harmony with custom.

The great martyr of our Church, when at the stake, is said to have held out his right hand into the flames and to have been heard exclaiming, till utterance was stifled, "This unworthy hand." This unworthy hand! Of whom or of what was that hand unworthy? Was it unworthy of Him who made it? Was it unworthy of him who bore it? Was it unworthy of the purposes for which it was made? Was it not, on the contrary, a too worthy hand? a hand worthy of a better usage than to be made, first, to sign a recantation of faith and, then, to be burned for having done so? a hand worthy of a better man? No one would have admitted this more readily than Cranmer. We may be sure that he would never have thought of proclaiming a hand or any of his members to be really unworthy of him. Rather would he have willingly confessed that he had fallen far short of the standard of excellence which the body presents; and in that excellence, we doubt not, he recognised an evidence of Divine workmanship. His meaning, therefore, has not been misunderstood. Nevertheless disparaging remarks respecting the body, and the use of the word " carnal" in the sense in which it is usually employed, have some tendency to excuse a shrinking from moral responsibilities on the ground of the weakness of the flesh. Let us remember that much of that weakness is of our own engendering, that a moral obliquity is the source of many of those physical infirmities which, we flatter ourselves, may cover our delinquencies, and which a sympathising humanity is wont, perhaps too often, to throw as a shield over offenders against the laws. In man, and in man alone of created beings, the physical and the moral grow up together and react upon one another; and the charge of a body thus capable of influencing and being influenced demands all our energies to prove ourselves worthy of it.

EXPLANATION OF WOOD-CUTS.

THE HUMAN FOOT.

Fig.	page	
7	25	The same bones as in preceding, with two connecting ligaments. A, the *plantar ligament*. B, ligament passing from the heel-bone F to the scaphoid bone E. D the *Astragalus*. C, one of two small bones, . called *sesamoid* bones, usually found at the ball of the great toe.
8	29	A foot, in an aggravated condition of "flat-foot." The sole is convex, and so is the inner margin of the foot. It represents also another common deformity, inasmuch as the great toe runs athwart the second toe, which is pressed almost out of sight.
9	37	Front view of the right *tibia*, or larger leg-bone.
10	38	Right *tibia* lying on a board. The inner, as well as the outer edge, of the upper end rests upon the board; but the inner edge of the lower end is turned away from the board. In other words, the bone is so twisted that, though the upper end lies flat upon the board, the lower end touches it only by its outer edge.
11	39	Figure sitting upon the heel to draw the bow. It is one of a beautiful series of statues in the Glyptothek at Munich. They adorned the pediments of a temple in Ægina, and are supposed to represent the noble actions of the Æacidæ.
12	42	Represents some of the muscles and tendons seen on the inner side of the leg and foot. A, *Gastrocnemius* and *Soleus* muscles. They are attached, above, to the thigh-bone and the leg-bones; below, by means of the *Tendo Achillis* (*a*) to the heelbone; they together form the calf-muscle. B, *Posterior tibial* muscle attached, above, to the tibia, below, by its tendon (*b*) to the scaphoid bone. D, process of the tibia called the *internal malleolus* or inner ankle. F, *Anterior tibial* muscle attached, above, to the front of the tibia, below,

H.

THE HUMAN HAND.

Fig.   page

flexor of the wrist. B, the *long palmar* muscle. C, the *ulnar flexor* of the wrist. D, the muscles of the "ball of the thumb." E, the *long supinator* muscle. F, the *long pronator*. G, the lower part of the *biceps* muscle. H, Cross *ligaments* binding the tendons in their places. (This and the two following figures are from Quain's *Anatomy*.)

66   139   View of the deep muscles and tendons on the palmar aspects of the forearm and hand. A, the *long flexor* of the thumb. B, some of the *flexors* of the fingers. C, the *Adductor* muscle of the thumb.

67   140   The *extensor* muscles and tendons of the wrist, thumb, and fingers seen on the back of the forearm and hand. A, A, A, the *abductors* and *adductors* of the fingers. B, B, the cross *ligament* which binds the tendons in their places.

68   146   Hand holding a cricket-ball, showing that the tips of the fingers and the thumb all reach the same level.

69   151   Diagram showing the distribution of the *median* (A) and *ulnar* (B) *nerves* in the hand.

70   165   Drawing of a magnified section through the skin of the palmar surface of the thumb, including three of the ridges seen on that surface. *a*, the outer or horny layer of the *cuticle; b*, the deeper layer of the same called "*rete mucosum;*" *c, c, c*, the *cutis*, with *papillæ* rising from its surface beneath the ridges and projecting into the rete mucosum ; *g, g*, grains of *fat* lying in the deeper part of the cutis and in the tissue beneath it. Between *f* and *f* are three *sweat-glands*, each composed of a tube rolled up into a ball or knot. The tubes (*h, h*) are seen ascending from them through the cutis

Fig.   page

and cuticle, and opening at the tops of the ridges. (From Kölliker's *Mikroskopishe Anatomie.*)

71   169   Section of skin still more magnified. *a*, Outer or horny stratum of cuticle; *b*, inner stratum of cuticle, or "rete mucosum;" *c*, papillary stratum of cutis; *d*, deeper or fibrous stratum of cutis. The curling tube rolled into a ball at the lower part is the sweat-gland. Its duct is seen ascending through the fibrous structure of the cutis, and presents the coiled appearance of a rope as it traverses the cuticle.

72   169   A few layers of the cuticle and rete mucosum of a Negro, showing the spots of dark pigment in the rete which give the black colour to the Negro's skin. (This and the preceding from Todd and Bowman's *Phys. Anatomy.*)

73   170   Section of a Corn and adjacent skin. *a*, the *cuticle;* *c*, the *cutis* with its *papillæ.* The cuticle is seen to be very thick, and the papillæ are somewhat enlarged in the corn.

74   170   Section of a Wart and adjacent skin. *a, cuticle; c*, the *cutis* with its *papillæ.* The latter are seen to be enlarged, or "hypertrophied," in the wart.

75   174   Vertical section, made lengthways, of a Nail raised from its bed, showing its connexion with the cuticle. *a, a, cuticle; d, d, nail.*

76   174   Similar section of a Nail lying in its bed of cutis. *a, cuticle; b, rete mucosum; c, cutis; d, nail.*

77   174   Section of the Cutis from which the nail, the cuticle, and the rete have been removed.

78   176   Transverse section of the Nail and Skin, made vertically. *a, a, cuticle; b, rete; c, c, cutis; d, d,* lines running through the cutis to the *papillæ; e, e, e,* lines running through the nail to the rete. (This and the three preceding from Kölliker.)

www.ingramcontent.com/pod-product-compliance
Lightning Source LLC
Chambersburg PA
CBHW030324270326
41926CB00010B/1491